## Advances in Mathematical Fluid Mechanics


**Series Editors**

Giovanni P. Galdi
School of Engineering
Department of Mechanical Engineering
University of Pittsburgh
3700 O'Hara Street
Pittsburgh, PA 15261
USA
e-mail: galdi@engrng.pitt.edu

John G. Heywood
Department of Mathematics
University of British Columbia
Vancouver BC
Canada V6T 1Y4
e-mail: heywood@math.ubc.ca

Rolf Rannacher
Institut für Angewandte Mathematik
Universität Heidelberg
Im Neuenheimer Feld 293/294
69120 Heidelberg
Germany
e-mail: rannacher@iwr.uni-heidelberg.de


*Advances in Mathematical Fluid Mechanics* is a forum for the publication of high quality monographs, or collections of works, on the mathematical theory of fluid mechanics, with special regards to the Navier-Stokes equations. Its mathematical aims and scope are similar to those of the *Journal of Mathematical Fluid Mechanics*. In particular, mathematical aspects of computational methods and of applications to science and engineering are welcome as an important part of the theory. So also are works in related areas of mathematics that have a direct bearing on fluid mechanics.

The monographs and collections of works published here may be written in a more expository style than is usual for research journals, with the intention of reaching a wide audience. Collections of review articles will also be sought from time to time.

# Contributions to Current Challenges in Mathematical Fluid Mechanics

Giovanni P. Galdi
John G. Heywood
Rolf Rannacher
Editors

Birkhäuser Verlag
Basel · Boston · Berlin

Editors:

Giovanni P. Galdi
School of Engineering
Department of Mechanical Engineering
University of Pittsburgh
3700 O'Hara Street
Pittsburgh, PA 15261
USA
e-mail: galdi@engrng.pitt.edu

John G. Heywood
Department of Mathematics
University of British Columbia
Vancouver BC
Canada V6T 1Y4
e-mail: heywood@math.ubc.ca

Rolf Rannacher
Institut für Angewandte Mathematik
Universität Heidelberg
Im Neuenheimer Feld 293/294
69120 Heidelberg
Germany
e-mail: rannacher@iwr.uni-heidelberg.de

2000 Mathematical Subject Classification 76D05, 35Q30, 76N10

A CIP catalogue record for this book is available from the
Library of Congress, Washington D.C., USA

Bibliographic information published by Die Deutsche Bibliothek
Die Deutsche Bibliothek lists this publication in the Deutsche Nationalbibliografie;
detailed bibliographic data is available in the Internet at <http://dnb.ddb.de>

ISBN 3-7643-7104-8 Birkhäuser Verlag, Basel – Boston – Berlin


Printed on acid-free paper produced from chlorine-free pulp. TCF ∞
Printed in Germany
ISBN 3-7643-7104-8

9 8 7 6 5 4 3 2 1

www.birkhauser.ch

# Contents

# Preface

This volume consists of five research articles, each dedicated to a significant topic in the mathematical theory of the Navier–Stokes equations, for compressible and incompressible fluids, and to related questions. All results given here are new and represent a noticeable contribution to the subject.

One of the most famous predictions of the Kolmogorov theory of turbulence is the so-called Kolmogorov–Obukhov five-thirds law. As is known, this law is heuristic and, to date, there is no rigorous justification. The article of A. Biryuk deals with the Cauchy problem for a multi-dimensional Burgers equation with periodic boundary conditions. Estimates in suitable norms for the corresponding solutions are derived for "large" Reynolds numbers, and their relation with the Kolmogorov–Obukhov law are discussed. Similar estimates are also obtained for the Navier–Stokes equation.

In the late sixties J. L. Lions introduced a "perturbation" of the Navier–Stokes equations in which he added in the linear momentum equation the hyper-dissipative term $(-\Delta)^{\beta}u$, $\beta \geq 5/4$, where $\Delta$ is the Laplace operator. This term is referred to as an "artificial" viscosity. Even though it is not physically motivated, artificial viscosity has proved a useful device in numerical simulations of the Navier–Stokes equations at high Reynolds numbers. The paper of of D. Chae and J. Lee investigates the global well-posedness of a modification of the Navier–Stokes equation similar to that introduced by Lions, but where now the original dissipative term $-\Delta u$ is replaced by $(-\Delta)^{\alpha}u$, $0 \leq \alpha < 5/4$. Existence, uniqueness and stability of solutions is proved in appropriate Besov spaces depending on the parameter $\alpha$.

Space averaged Navier–Stokes equations are the basic equations for large eddy simulation of turbulent flows. In deriving these equations it is tacitly understood that differentiation and averaging operations can be interchanged. Actually, this procedure introduces a "commutation error" term that is typically ignored. The main objective in the paper of A. Dunca, V. John and W. L. Layton is to furnish a characterization of this term to be neglected. The authors go on to provide a justification for neglecting this term if and only if the Cauchy stress vector of the underlying flow is identically zero on the boundary of the domain. In other words, neglecting the commutation error is reasonable only for flows in which the boundary exerts no influence on the flow.

Since the appearence of the paper of J. G. Heywood in the mid-seventies, the problem of a flow through an aperture (the "aperture domain" problem) has attracted the attention of many researchers. But even now, a number of basic questions remain unresolved.The article of T. Hishida provides a further, significant contribution. Specifically, the author proves $L^q - L^r$ estimates for the Stokes semigroup in an aperture domain of $\mathbb{R}^n$, $n \geq 3$. These estimates are then used to show the existence, uniqueness and asymptotic behavior in time of strong solutions of the Navier–Stokes equations having "small" initial data in $L^n$ and zero flux through the aperture.

The mathematical analysis of the well-posedness of the Navier–Stokes equations in the case of a compressible viscous fluid is a relatively new branch of mathematical fluid mechanics, its first contribution dating back to a paper of J. Nash in the early sixties. Many problems remain to be solved in this area, despite the significant contributions of many mathematicians. In particular, there remain very interesting problems concerning steady flow in an exterior domain, especially regarding the asymptotic behavior of solutions. This latter problem is analyzed in the paper of T. Leonavičienė and K. Pileckas, in the case when the velocity of the fluid is zero at large distances and the body force is the sum of an "arbitrary large" potential term and a "small" non-potential term.

We would like to express our warm thanks to Professors H. Beirão da Veiga, A. Fursikov and Y. Giga who recommended the publication of these articles.

Giovanni P. Galdi John G. Heywood Rolf Rannacher

Advances in Mathematical Fluid Mechanics, 1–30


# On Multidimensional Burgers Type Equations with Small Viscosity

Andrei Biryuk

**Abstract.** We consider the Cauchy problem for a multidimensional Burgers type equation with periodic boundary conditions. We obtain upper and lower bounds for derivatives of solutions for this equation in terms of powers of the viscosity and discuss how these estimates relate to the Kolmogorov–Obukhov spectral law. Next we use the estimates obtained to get certain bounds for derivatives of solutions of the Navier–Stokes system.

**Mathematics Subject Classification (2000).** 35B10, 35A30, 76D05.

**Keywords.** Kolmogorov–Obukhov spectral law, bounds for derivatives, degenerate state.

## 1. Introduction

We study the dynamics of $m$-dimensional vector field $\boldsymbol{u} = \boldsymbol{u}(t, \boldsymbol{x})$ on the $n$-dimensional torus $\mathbb{T}^n = {}^{\mathbb{R}^n}\!/_{(\ell\mathbb{Z})^n}$ described by the equation

$$\partial_t \boldsymbol{u} + \nabla_{\boldsymbol{f}(\boldsymbol{u})} \boldsymbol{u} = \nu \Delta \boldsymbol{u} + \boldsymbol{h}(t, \boldsymbol{x}) \,. \tag{1.1}$$

Here $\nu$ is a positive parameter ("the viscosity"), $\boldsymbol{f} : \mathbb{R}^m \to \mathbb{R}^n$ is a smooth map, $\boldsymbol{h}$ is a smooth forcing term and $\nabla_{\boldsymbol{f}(\boldsymbol{u})}$ is the derivative along the vector $\boldsymbol{f}(\boldsymbol{u})$, i.e., $\nabla_{\boldsymbol{f}(\boldsymbol{u})} \boldsymbol{u} = (\boldsymbol{f}(\boldsymbol{u}) \cdot \nabla) \boldsymbol{u}$.

If $m = n$ and $\boldsymbol{f}(\boldsymbol{u}) = \boldsymbol{u}$, we have the usual forced Burgers equation. In a potential case (i.e., if the initial state $\boldsymbol{u}_0(\boldsymbol{x}) = \boldsymbol{u}(0, \boldsymbol{x})$ and the field $\boldsymbol{h}$ are gradients of some functions) this equation can be reduced to a linear parabolic equation.

As it is shown in [1], [12], appropriate bounds for derivatives imply estimates for averaged spectral characteristics of the flow. The purpose of this work is to obtain such bounds for solutions of the Cauchy problem for the generalised $m$-$n$ multidimensional Burgers equation (1.1).

We describe notations used in this article. If $\boldsymbol{v}$ is a vector in $\mathbb{R}^s$, $\mathbb{Z}^s$ or $\mathbb{C}^s$, then $|\boldsymbol{v}|$ denotes its Euclidean (Hermitian) norm $|\boldsymbol{v}|^2 = \sum_{i=1}^{s} |v_i|^2$. If we have to stress the dimension, we denote the norm in $\mathbb{R}^s$ as $|\boldsymbol{v}|_{\mathbb{R}^s}$, etc. By $B(r)$ we denote the ball of radius $r$ centered at the origin. If $A : \mathbb{R}^{s_1} \to \mathbb{R}^{s_2}$ is a linear map then $\|A\|$ denotes the operator-norm of this map associated with the Euclidean norms $|\cdot|$ on $\mathbb{R}^{s_1}$ and $\mathbb{R}^{s_2}$. If $\boldsymbol{v} = \boldsymbol{v}(\boldsymbol{x})$, then we write

$$|\boldsymbol{v}| = \sup_{\boldsymbol{x}} |\boldsymbol{v}(\boldsymbol{x})| = \sup_{\boldsymbol{x}} \left(\sum |v_i(\boldsymbol{x})|^2\right)^{1/2}. \tag{1.2}$$

Sometimes we will denote this norm by $|\cdot|_{L_\infty}$. If $\boldsymbol{v} = \boldsymbol{v}(t,\boldsymbol{x})$, then $|\boldsymbol{v}| = \sup_{\boldsymbol{x}} |\boldsymbol{v}(t,\boldsymbol{x})|$ is a function of $t$. For a multi-index $\alpha$ we denote $|\alpha| = \sum |\alpha_i|$.

We set

$$H(t) = \int_0^t \sup_{\boldsymbol{x}\in\mathbb{T}^n} |\boldsymbol{h}(\tau,\boldsymbol{x})|_{\mathbb{R}^m}\, d\tau\,. \tag{1.3}$$

We also denote

$$[\boldsymbol{f}]_{C^k(r)} = \max_{|\beta|=k} \sup_{\{\boldsymbol{u}\in\mathbb{R}^m : |\boldsymbol{u}|\leqslant r\}} \left(\sum_{j=1}^n \left|\frac{\partial^k}{\partial \boldsymbol{u}^\beta} f_j\right|^2\right)^{1/2}, \tag{1.4}$$

and

$$\|\boldsymbol{u}\|_k^2 = \int_{\mathbb{T}^n} \sum_{i=1}^m u_i(-\Delta)^k u_i d\boldsymbol{x} = \sum_{i=1}^m \sum_{|\alpha|=k} \binom{|\alpha|}{\alpha} \left|D^\alpha u_i\right|^2_{L_2(\mathbb{T}^n)}$$

$$= \sum_{i=1}^m \sum_{j_1,..,j_k=1}^n \left|\frac{\partial^k u_i}{\partial x_{j_1}\cdots\partial x_{j_k}}\right|^2_{L_2(\mathbb{T}^n)}. \tag{1.5}$$

Here $k \geqslant 0$ is an integer and $\binom{|\alpha|}{\alpha} = \binom{|\alpha|}{\alpha_1,\dots,\alpha_n} = \frac{(\alpha_1+\dots+\alpha_n)!}{\alpha_1!\cdots\alpha_n!}$ are coefficients in the generalised binomial expansion $(x_1 + x_2 + \ldots + x_n)^k = \sum_{|\alpha|=k} \binom{k}{\alpha} \boldsymbol{x}^\alpha$. If $\boldsymbol{u} = \boldsymbol{u}(t,\boldsymbol{x})$, then $\|\boldsymbol{u}\|_k = \|\boldsymbol{u}(t,\cdot)\|_k$ is a function of $t$.

Our main results are stated in the following two theorems, where $\boldsymbol{u}(t,\boldsymbol{x})$, $t \geqslant 0$, is any smooth solution for the equation (1.1):

**Theorem 1.** *For any $k \geqslant 0$, $t \geqslant 0$, and $\nu > 0$ we have*

$$\|\boldsymbol{u}(t,\cdot)\|_k \leqslant R_k(t) \max\left\{\tfrac{1}{\nu^k}, \tfrac{\|\boldsymbol{u}_0\|_k}{R_k(0)}, \tfrac{\sup_{[0,t]}\|\boldsymbol{h}\|_{k-1}}{R_k(0)}\right\}. \tag{1.6}$$

*Here $\|\boldsymbol{h}\|_{-1} = 0$, $R_0(t) = (|\boldsymbol{u}_0| + H(t))\ell^{n/2}$ and*

$$R_k(t) = \left(1 + C_{k,m,n} \max_{s=0\ldots k-1}\{[\boldsymbol{f}]_{C^s(|\boldsymbol{u}_0|+H(t))}\,(|\boldsymbol{u}_0| + H(t))^s\}\right)^k (|\boldsymbol{u}_0| + H(t))\ell^{n/2},$$

*where the constant $C_{k,m,n}$ depends on $k$, $m$, $n$ only.*

**Definition 1.** The vector field $\boldsymbol{u}_0$ is degenerate with respect to equation (1.1) if the matrix $\frac{\partial \boldsymbol{f}(\boldsymbol{u}_0)}{\partial \boldsymbol{x}}$ (this is an $n \times n$ matrix, which depends on $\boldsymbol{x}$) is everywhere nilpotent, i.e., for each point $\boldsymbol{x}$ some power of this matrix is equal to 0.

**Theorem 2.** *Suppose that the initial state $\boldsymbol{u}_0$ is a non-degenerate vector field. Then there exist $\nu$-independent positive real constants $T$, $c$ and $r_2$, $r_3$, $r_4$, ... such that:*
*If $H(T) < \frac{c}{2}$ then $\forall \nu > 0$ and $\forall k \geqslant 2$, we have:*

$$\max_{j=1\ldots n} \frac{1}{T} \int_0^T \sup_{\boldsymbol{x}\in\mathbb{T}^n} \left| \frac{\partial^k \boldsymbol{u}}{\partial {x_j}^k}(t,\boldsymbol{x}) \right|_{\mathbb{R}^m} dt \geqslant \frac{r_k}{\nu^{k/2}}. \tag{1.7}$$

The constants $T$, $c$, $r_2$, $r_3$,... depend on the non-degeneracy of the initial state $\boldsymbol{u}_0$ in quite complicated way (see (3.11), (3.16), (3.19), and (3.33)). The nearer is $\boldsymbol{u}_0$ to set of degenerate vector functions, the bigger is $T$ and the smaller are $c$, $r_2$, $r_3$,.... If $\boldsymbol{u}_0 = \lambda \boldsymbol{v_0}$ and $\lambda \to 0$, then $T \propto \lambda^{-1}$, $r_k \propto \lambda^{k/2}$, $c$ does not depend on $\lambda$.

In Section 3 we give an example of a degenerate non-constant initial state for which derivatives of the solution are bounded by $\nu$-independent constants for all $t \geqslant 0$. Moreover, for the two dimensional case ($m = n = 2$) and for $\boldsymbol{f}(\boldsymbol{u}) = \boldsymbol{u}$, $\boldsymbol{h} \equiv 0$ we show that any solution with a degenerate initial state retains bounded derivatives. This fact is based on a result from the classical geometry due to Pogorelov–Hartman–Nirenberg, known as the "Cylinder Theorem". In Section 3 we show that in the case $m = n$ and $\boldsymbol{f}(\boldsymbol{u}) = \boldsymbol{u}$, any non-constant *potential* initial state is non-degenerate.

The exponents of viscosity $\nu$ in inequalities (1.6) and (1.7) are not sharp. In the one-dimensional case ($m = n = 1$) sharp values for the exponents can be obtained. Namely, it is shown in [1] that for $k \geqslant 1$ we have

$$\|\boldsymbol{u}\|_k \leqslant C_k \left(\tfrac{1}{\nu}\right)^{k-1/2}, \quad \left(\frac{1}{T}\int_0^T \|\boldsymbol{u}\|_k^2 dt\right)^{1/2} \geqslant c_k \left(\tfrac{1}{\nu}\right)^{k-1/2}.$$

As a consequence of these inequalities one can get bounds for magnitudes of the derivatives:

$$\left|\frac{d^k u}{dx^k}\right|_{L_\infty} \leqslant \frac{C'_k}{\nu^k}, \quad \frac{1}{T}\int_0^T \left|\frac{d^k u}{dx^k}\right|_{L_\infty} dt \geqslant \frac{c'_k}{\nu^k}.$$

The first inequality for $k = 0$ follows by the maximum principle and for $k \geqslant 1$ – by the inequality $|v|_{L_\infty} \leqslant |v|_{L_2}^{1/2} |v_x|_{L_2}^{1/2}$ which holds for any periodic function $v$ with zero meanvalue, see [1], Sect. 3. To derive the second inequality (see [1] and formula (3.7) there) we use the well-known fact that for periodic solutions of 1D Burgers-type equations the quantity $\left|\frac{du}{dx}\right|_{L_1}$ is bounded uniformly in $t$ (see e.g. the appendix in [1]). Then for $k = 1$ the second estimate follows by the Hölder inequality $|u_x|_{L_2}^2 \leqslant |u_x|_{L_\infty} |u_x|_{L_1}$, while for $k > 1$ it follows by interpolation with the upper bound for $k = 0$.

This article is organised as follows. In Section 2 we prove the upper estimates (1.6) (theorem 1). Section 3 is devoted to proving the lower bounds (1.7) (theorem 2). In Section 4 we obtain some results on behaviour of Fourier coefficients of solutions that can be extracted from the bounds (1.6) and (1.7). Assuming that there is a Kolmogorov–Obukhov type spectral asymptotics for the Fourier coefficients of solutions of (1.1), we get bounds for the exponents of the spectral law and for the

Kolmogorov dissipation scale. In Section 5, treating the Navier–Stokes system as a partial case of (1.1), we derive lower bounds for derivatives of its solutions.

The author is grateful to Professor S. Kuksin for constant attention to this work.

## 2. Upper estimates

In this section we prove Theorem 1. The componentwise representation of (1.1) is

$$\frac{\partial}{\partial t}u_i + \sum_{j=1}^{n} f_j(u_1,\dots,u_m)\frac{\partial u_i}{\partial x_j} = \nu\Delta u_i + h_i(t,x_1,\dots,x_n)\,, \tag{2.1}$$

where $i = 1,\dots,m$.

**Lemma 1.** *Let $T$ and $\nu$ be any positive numbers. Let $v = v(t,\boldsymbol{x})$ be a continuous function on $[0,T]\times\mathbb{T}^n$ with the continuous derivatives $v_t$, $v_{x_j}$ and $v_{x_jx_j}$ for any $j = 1,\dots,n$. Let $V_j = V_j(t,\boldsymbol{x})$, $j = 1,\dots,n$ and $g = g(t,\boldsymbol{x})$ be continuous functions on $[0,T]\times\mathbb{T}^n$. Suppose that on $[0,T]\times\mathbb{T}^n$ we have the following partial differential inequality:*

$$v_t + \sum_{j=1}^{n} V_j\frac{\partial v}{\partial x_j} \leqslant \nu\Delta v + g(t,x_1,\dots,x_n).$$

*Then for any $(t,\boldsymbol{x}) \in [0,T]\times\mathbb{T}^n$ we have*

$$v(t,\boldsymbol{x}) \leqslant \max_{\boldsymbol{y}\in\mathbb{T}^n} v(0,\boldsymbol{y}) + \int_0^t \max_{\boldsymbol{y}\in\mathbb{T}^n} g(\tau,\boldsymbol{y})d\tau.$$

*Proof.* Making the substitution $v(t,\boldsymbol{x}) = \tilde{v}(t,\boldsymbol{x}) + q(t)$, where $q{:}[0,T]\to\mathbb{R}$ is a function such that $q'(t) = \max_{\boldsymbol{y}\in\mathbb{T}^n} g(t,\boldsymbol{y})$, we reduce this lemma to the case $g \equiv 0$. Now the statement of the lemma becomes a classic maximum principle, see e.g. [4]. □

Applying this lemma for $v(t,\boldsymbol{x}) = \sum a_i u_i(t,\boldsymbol{x})$ and $g(t,\boldsymbol{x}) = \sum a_i h_i(t,\boldsymbol{x})$ with appropriate unit vector $\boldsymbol{a}\in\mathbb{R}^m$ we obtain

$$|\boldsymbol{u}(t,\cdot)| \leqslant |\boldsymbol{u}_0| + H(t)\,. \tag{2.2}$$

Here the norm $|\cdot|$ is defined by (1.2) and $H(t)$ is defined by (1.3).

Since $\|\boldsymbol{u}\|_0 \leqslant \ell^{n/2}\,|\boldsymbol{u}|$, we have

$$\|\boldsymbol{u}(t,\cdot)\|_0 \leqslant \ell^{n/2}(|\boldsymbol{u}_0| + H(t))\,. \tag{2.3}$$

This proves (1.6) for $k = 0$. Next, we multiply (2.1) by $(-\Delta)^k u_i$, take the sum over $i = 1,\dots,m$, and integrate over the period (over the torus):

$$\tfrac{1}{2}\tfrac{d}{dt}\|\boldsymbol{u}\|_k^2 - b(\boldsymbol{f},\boldsymbol{u},(-\Delta)^k\boldsymbol{u}) = -\nu\|\boldsymbol{u}\|_{k+1}^2 + \Upsilon_2^k\,.$$

Here we denote

$$b(\boldsymbol{f},\boldsymbol{u},\boldsymbol{v}) = -\int_{\mathbb{T}^n} \sum_{\substack{j=1\dots n\\ i=1\dots m}} f_j\frac{\partial u_i}{\partial x_j}v_i d\boldsymbol{x}\,. \tag{2.4}$$

and

$$\Upsilon_2^k = \int_{\mathbb{T}^n} \sum_{i=1\dots m} h_i(-\Delta)^k u_i d\boldsymbol{x}\,.$$

**Lemma 2.** *For the functional b introduced above we have*

$$b(\boldsymbol{f}(\boldsymbol{u}),\boldsymbol{u},\boldsymbol{u}) \leqslant [\boldsymbol{f}]_{C^0(|\boldsymbol{u}|)} \|\boldsymbol{u}\|_1 \|\boldsymbol{u}\|_0\,, \tag{2.5}$$

$$b(\boldsymbol{f}(\boldsymbol{u}),\boldsymbol{u},(-\Delta)\boldsymbol{u}) \leqslant [\boldsymbol{f}]_{C^0(|\boldsymbol{u}|)} \|\boldsymbol{u}\|_1 \|\boldsymbol{u}\|_2\,, \tag{2.6}$$

*and for any $k \geqslant 2$ we have*

$$b(\boldsymbol{f}(\boldsymbol{u}),\boldsymbol{u},(-\Delta)^k\boldsymbol{u}) \leqslant C_{k,m,n} \max_{s=0,\dots,k-1}\{[\boldsymbol{f}]_{C^s(|\boldsymbol{u}|)} |\boldsymbol{u}|^s_{L_\infty}\} \|\boldsymbol{u}\|_k \|\boldsymbol{u}\|_{k+1}\,. \tag{2.7}$$

*Proof.* First we prove a general inequality on $b(\cdot,\cdot,\cdot)$:

$$|b(\boldsymbol{f},\boldsymbol{u},\boldsymbol{v})| \leqslant |\boldsymbol{f}| \|\boldsymbol{u}\|_1 \|\boldsymbol{v}\|_0\,, \tag{2.8}$$

where $|\boldsymbol{f}| = \sup\bigl(\sum_{j=1}^n f_j^2\bigr)^{1/2}$ and the norms $\|\cdot\|_s$ are defined in (1.5). Up to a constant factor this inequality is obvious. Below we show that for the chosen norm the constant is equal to 1. By the definition (2.4) of $b(\cdot,\cdot,\cdot)$ and the Cauchy–Schwartz inequality we have

$$|b(\boldsymbol{f},\boldsymbol{u},\boldsymbol{v})| \leqslant \int \Bigl(\sum_{j=1}^n f_j^2\Bigr)^{1/2} \Bigl(\sum_{j=1}^n\Bigl(\sum_{i=1}^m \tfrac{\partial u_i}{\partial x_j} v_i\Bigr)^2\Bigr)^{1/2} dx$$

now we again use the Cauchy–Schwartz inequality $(\sum a_i b_i)^2 \leqslant (\sum a_i^2)(\sum b_i^2)$ to continue as follows:

$$\leqslant |\boldsymbol{f}| \int \Bigl(\sum_{j=1}^n\Bigl(\sum_{i=1}^m \bigl(\tfrac{\partial u_i}{\partial x_j}\bigr)^2\Bigr)\Bigl(\sum_{i=1}^m v_i^2\Bigr)\Bigr)^{1/2} dx$$

$$= |\boldsymbol{f}| \int \Bigl(\sum_{j=1}^n\sum_{i=1}^m \bigl(\tfrac{\partial u_i}{\partial x_j}\bigr)^2\Bigr)^{1/2}\Bigl(\sum_{i=1}^m v_i^2\Bigr)^{1/2} dx$$

$$\leqslant |\boldsymbol{f}| \Bigl(\int \sum_{j=1}^n\sum_{i=1}^m \bigl(\tfrac{\partial u_i}{\partial x_j}\bigr)^2 d\boldsymbol{x}\Bigr)^{1/2}\Bigl(\int \sum_{i=1}^m v_i^2 d\boldsymbol{x}\Bigr)^{1/2} = |\boldsymbol{f}| \|\boldsymbol{u}\|_1 \|\boldsymbol{v}\|_0\,.$$

The inequality (2.8) and therefore (2.5) are proved. Using $\|\Delta\boldsymbol{u}\|_0 = \|\boldsymbol{u}\|_2$ we arrive at (2.6).

Consider the case $k \geqslant 2$. By (2.4) we have

$$b(\boldsymbol{f},\boldsymbol{u},(-\Delta)^k\boldsymbol{u}) = (-1)^{k-1}\int \sum_{j_0,\dots,j_k=1}^n \sum_{i=1}^m f_{j_0}(u_1,\dots,u_m)\frac{\partial u_i}{\partial x_{j_0}}\frac{\partial^2}{\partial x_{j_1}^2}\cdots\frac{\partial^2}{\partial x_{j_k}^2}u_i d\boldsymbol{x}.$$

Integrating by parts $k-1$ times we obtain

$$b(\boldsymbol{f},\boldsymbol{u},(-\Delta)^k\boldsymbol{u})$$

$$= \int \sum_{j_0,\dots,j_k=1}^n \sum_{i=1}^m \frac{\partial}{\partial x_{j_1}}\cdots\frac{\partial}{\partial x_{j_{k-1}}}\Bigl(f_{j_0}(u_1,\dots,u_m)\frac{\partial u_i}{\partial x_{j_0}}\Bigr)\frac{\partial}{\partial x_{j_1}}\cdots\frac{\partial}{\partial x_{j_{k-1}}}\frac{\partial^2}{\partial x_{j_k}^2}u_i\, d\boldsymbol{x}.$$

Using the identity

$$\|\boldsymbol{u}\|_{k+1}^2 = \int \sum_{j_1,\dots,j_{k-1}=1,\dots,n} \sum_{i=1}^{m} \Big(\sum_{j_k=1}^{n} \frac{\partial}{\partial x_{j_1}} \cdots \frac{\partial}{\partial x_{j_{k-1}}} \frac{\partial^2}{\partial x_{j_k}^2} u_i\Big)^2 d\boldsymbol{x}$$

we get

$$|b(\boldsymbol{f},\boldsymbol{u},(-\Delta)^k\boldsymbol{u})|$$
$$\leqslant \Big(\int \sum_{j_0,\dots,j_{k-1}=1}^{n} \sum_{i=1}^{m} \Big(\frac{\partial}{\partial x_{j_1}} \cdots \frac{\partial}{\partial x_{j_{k-1}}} \Big(f_{j_0}(u_1,\dots,u_m)\frac{\partial u_i}{\partial x_{j_0}}\Big)\Big)^2 d\boldsymbol{x}\Big)^{1/2} \|\boldsymbol{u}\|_{k+1}$$

Now to prove the lemma it suffices to verify the inequality

$$\int \Big|\frac{\partial}{\partial x_{j_1}} \cdots \frac{\partial}{\partial x_{j_{k-1}}} \Big(f_{j_0}(u_1,\dots,u_m)\frac{\partial u_i}{\partial x_{j_0}}\Big)\Big| dx$$
$$\leqslant C'_{k,m,n} \max_{s=0,\dots,k-1} \{[\boldsymbol{f}]_{C^s(|\boldsymbol{u}|)} |\boldsymbol{u}|_{L_\infty}^s\} \|\boldsymbol{u}\|_k . \quad (2.9)$$

(Indeed, (2.9) implies (2.7) with $C_{k,m,n} = (mn^{k-1})^{1/2}C'_{k,m,n}$.) Expanding the brackets in (2.9) we get no more than $(m+1)(m+2)\cdots(m+k-1)$ terms of the form

$$\int_{\mathbb{T}^n} D^\beta_{\boldsymbol{u}} f_{j_0} D^{\alpha_0}_{\boldsymbol{x}} u_i D^{\alpha_1}_{\boldsymbol{x}} u_{i_1} \cdots D^{\alpha_{|\beta|}}_{\boldsymbol{x}} u_{i_{|\beta|}} d\boldsymbol{x}\,,$$

where $|\alpha_0| + |\alpha_1| + \cdots + |\alpha_{|\beta|}| = k$ and the indexes $i_s$ (where $s = 1,\dots,|\beta|$) vary between 1 and $m$. The modulus of this integral is not bigger than

$$\mathfrak{A} = |D^\beta_{\boldsymbol{u}} f_{j_0}|_{L_\infty(B(|\boldsymbol{u}|))} |D^{\alpha_0}_{\boldsymbol{x}} u_i|_{L_{\frac{2k}{|\alpha_0|}}} |D^{\alpha_1}_{\boldsymbol{x}} u_{i_1}|_{L_{\frac{2k}{|\alpha_1|}}} \cdots |D^{\alpha_{|\beta|}}_{\boldsymbol{x}} u_{i_{|\beta|}}|_{L_{\frac{2k}{|\alpha_{|\beta|}|}}} .$$

Here $B(r)$ denotes the ball in $\mathbb{R}^m$ of radius $r$ in the Euclidean norm, centered at the origin. Using the Gagliardo–Nirenberg inequality (see [9], pp. 106–107)

$$|D^{\alpha_s}_{\boldsymbol{x}} u_{i_s}|_{L_{\frac{2k}{|\alpha_s|}}} \leqslant 4^{|\alpha_s|(k-|\alpha_s|)} |u_{i_s}|_{L_\infty}^{1-\frac{|\alpha_s|}{k}} \|u_{i_s}\|_k^{\frac{|\alpha_s|}{k}},$$

and the inequality $\sum_{s=0}^{|\beta|}(|\alpha_s|k - |\alpha_s|^2) \leqslant \sum_{s=0}^{|\beta|}(|\alpha_s|k - |\alpha_s|) = k^2 - k$ we obtain

$$\mathfrak{A} \leqslant 4^{k^2-k} |D^\beta_{\boldsymbol{u}} f_j|_{L_\infty(B(|\boldsymbol{u}|))} |\boldsymbol{u}|_{L_\infty}^{|\beta|} \|\boldsymbol{u}\|_k .$$

Now using the fact that left hand side of (2.9) $\leqslant (m+1)\cdots(m+k-1)\max\{\mathfrak{A}\}$, we arrive at (2.9) with $C'_{k,m,n} = 4^{k^2-k}(m+1)(m+2)\cdots(m+k-1)$. □

**Corollary 1.** *For $k \geqslant 1$ we have:*

$$b(\boldsymbol{f}(\boldsymbol{u}),\boldsymbol{u},(-\Delta)^k\boldsymbol{u}) \leqslant B_k(t)\,\|\boldsymbol{u}\|_k\,\|\boldsymbol{u}\|_{k+1}\,,$$

*where* $B_k(t) = C_{k,m,n} \max_{s=0\dots k-1} \{[\boldsymbol{f}]_{C^s(|\boldsymbol{u}_0|+H(t))}(|\boldsymbol{u}_0| + H(t))^s\}$.

Integrating $\Upsilon_2^k$ by parts, we obtain $\Upsilon_2^k \leqslant \|\boldsymbol{h}\|_{k-1} \|\boldsymbol{u}\|_{k+1}$, So we have

$$\frac{1}{2}\frac{d}{dt}\|\boldsymbol{u}\|_k^2 \leqslant \|\boldsymbol{u}\|_{k+1}\left(-\nu\|\boldsymbol{u}\|_{k+1} + B_k(t)\|\boldsymbol{u}\|_k + \|\boldsymbol{h}\|_{k-1}\right).$$

Now using the interpolation inequality in the form $\|\boldsymbol{u}\|_{k+1} \geqslant \|\boldsymbol{u}\|_k \left(\frac{\|\boldsymbol{u}\|_k}{\|\boldsymbol{u}\|_0}\right)^{1/k}$, we have

$$\frac{1}{2}\frac{d}{dt}\|\boldsymbol{u}\|_k^2 \leqslant \|\boldsymbol{u}\|_{k+1}\left(\|\boldsymbol{u}\|_k\left(-\nu\left(\frac{\|\boldsymbol{u}\|_k}{\|\boldsymbol{u}\|_0}\right)^{1/k} + B_k(t)\right) + \|\boldsymbol{h}\|_{k-1}\right).$$

It follows from this relation that

$$\text{if } \|\boldsymbol{u}\|_k > \frac{(B_k(t)+1)^k\|\boldsymbol{u}\|_0}{\nu^k} \text{ and } \|\boldsymbol{u}\|_k > \|\boldsymbol{h}\|_{k-1}, \text{ then } \|\boldsymbol{u}\|_k \text{ is decreasing.} \tag{2.10}$$

We denote the right hand side of (1.6) by $F_k(t)$. It is clear from the definition of the function $F_k$ that

$$\|\boldsymbol{u}(0,\cdot)\|_k \leqslant F_k(0).$$

Using (2.3) we see that if $\|\boldsymbol{u}\|_k > F_k(t)$ then $\|\boldsymbol{u}\|_k$ is decreasing by argument (2.10). Since $F_k(t)$ is a non-decreasing function we obtain that $\|\boldsymbol{u}\|_k$ never can be greater than $F_k(t)$. We arrive at (1.6). Theorem 1 is proven.

## 3. Lower estimates

In this section we prove Theorem 2. Throughout this section we use standard facts from linear algebra about linear transformations. For the convenience of the reader we very briefly outline the proofs. See reference [7], for an elegant, coordinate free presentation. We start from brief discussing of the notion of the degeneracy of a vector field.

### 3.1. Degeneracy condition

Using the fact that an $n \times n$ matrix $A$ is nilpotent iff $A^n = 0$, we can give a definition of degeneracy that is equivalent to the previous one, but more robust.

**Definition 2.** The vector field $\boldsymbol{u}_0$ is degenerate iff $\left(\frac{\partial \boldsymbol{f}(\boldsymbol{u}_0(\boldsymbol{x}))}{\partial \boldsymbol{x}}\right)^n \equiv 0.$

Let $\tilde{\boldsymbol{f}}(\boldsymbol{x}) = \boldsymbol{f}(\boldsymbol{u}_0(\boldsymbol{x}))$. Consider the characteristic polynomial of the matrix $\frac{\partial \tilde{\boldsymbol{f}}}{\partial \boldsymbol{x}}$:

$$\chi_{\boldsymbol{x}}(\lambda) = \det\left(\frac{\partial \tilde{\boldsymbol{f}}}{\partial \boldsymbol{x}} - \lambda \mathbf{1}\right) = (-\lambda)^n + (-\lambda)^{n-1}I_1(\boldsymbol{x}) + \cdots + I_n(\boldsymbol{x}).$$

Expanding the determinant, we obtain

$$I_k(\boldsymbol{x}) = \sum_{1 \leqslant i_1 < i_2 < \cdots < i_k \leqslant n} \det \begin{pmatrix} \frac{\partial \tilde{f}_{i_1}}{\partial x_{i_1}} & \cdots & \frac{\partial \tilde{f}_{i_1}}{\partial x_{i_k}} \\ \vdots & \ddots & \vdots \\ \frac{\partial \tilde{f}_{i_k}}{\partial x_{i_1}} & \cdots & \frac{\partial \tilde{f}_{i_k}}{\partial x_{i_k}} \end{pmatrix}. \tag{3.1}$$

Using the Jordan form of the matrix we see that if $\frac{\partial \tilde{f}_i}{\partial x_j}(\boldsymbol{x})$ is nilpotent then all $I_j(\boldsymbol{x})$ are zero numbers. From the Hamilton–Cayley identity (any matrix is a root of its characteristic polynomial) we get the converse. Thus we have that the matrix $\frac{\partial \tilde{f}_i}{\partial x_j}(\boldsymbol{x})$ is nilpotent iff $I_1(\boldsymbol{x}) = \cdots = I_n(\boldsymbol{x}) = 0$. We got another equivalent definition of degeneracy which we will use subsequently:

**Definition 3.** The vector field $\boldsymbol{u}_0$ is degenerate iff $I_k(\boldsymbol{x}) \equiv 0$ for all $k \in \{1, \ldots, n\}$.

We also note that if $m < n$, then the matrix $\frac{\partial \tilde{f}_i}{\partial x_j}$ has rank $\leqslant m$, so for $k \in [m+1, n]$ we have $I_k(\boldsymbol{x}) \equiv 0$.

**Lemma 3.** *For each* $k = 1, \ldots, n$ *we have* $\int_{\mathbb{T}^n} I_k(\boldsymbol{x}) d\boldsymbol{x} = 0$ .

*Proof.* We need to show that

$$\int_{\mathbb{T}^n} \det \left( \frac{\partial \tilde{f}_i}{\partial x_j}(\boldsymbol{x}) + \lambda \mathbf{1} \right) d\boldsymbol{x} \equiv (\ell \lambda)^n. \tag{3.2}$$

Since both the left hand side and the right hand side are polynomials in $\lambda$, it is sufficient to prove this equality for all integer $\lambda$. We write $\frac{\partial \tilde{f}_i}{\partial x_j}(\boldsymbol{x}) + \lambda \mathbf{1} = \frac{\partial \boldsymbol{\Psi}}{\partial \boldsymbol{x}}$, where the vector valued function $\boldsymbol{\Psi}$ is defined by the formula $\boldsymbol{\Psi}(\boldsymbol{x}) = \tilde{\boldsymbol{f}}(\boldsymbol{x}) + \lambda \boldsymbol{x}$. Since $\lambda$ is an integer, then this function defines a map from $\mathbb{T}^n$ to $\mathbb{T}^n$. Since $\boldsymbol{\Psi}$ is homotopic to the map $\boldsymbol{x} \mapsto \lambda \boldsymbol{x}$ on torus $\mathbb{T}^n$ (a homotopy is given by $\boldsymbol{\Psi}_t(\boldsymbol{x}) = t\tilde{\boldsymbol{f}}(\boldsymbol{x}) + \lambda \boldsymbol{x}$), we have $\deg \boldsymbol{\Psi} = \deg\{\boldsymbol{x} \mapsto \lambda \boldsymbol{x}\}$ and hence $\deg \boldsymbol{\Psi} = \lambda^n$. Using the formula

$$\int_{\mathbb{T}^n} \det \frac{\partial \boldsymbol{\Psi}}{\partial \boldsymbol{x}} d\boldsymbol{x} = \deg \boldsymbol{\Psi} \int_{\mathbb{T}^n} d\boldsymbol{x}$$

(see [3], II, chpt. 3) we arrive at (3.2) (since $\int_{\mathbb{T}^n} d\boldsymbol{x} = \ell^n$). □

It follows that any potential degenerate initial state is constant. Indeed, if $\boldsymbol{f}(\boldsymbol{u}_0) = \nabla U$ then the function $U : \mathbb{R}^n \to \mathbb{R}$ necessarily has no more than linear growth (because $\boldsymbol{f}(\boldsymbol{u}_0(\boldsymbol{x}))$ is periodic) and is also a harmonic function (because $\Delta U(\boldsymbol{x}) = \operatorname{div} \boldsymbol{f}(\boldsymbol{u}_0(\boldsymbol{x})) = I_1(\boldsymbol{x}) \equiv 0$); so $U(\boldsymbol{x}) = B\boldsymbol{x} + \boldsymbol{c}$ where $B$ and $\boldsymbol{c}$ are a constant matrix and a constant vector, respectively.

Consider the case $n = 2$, i.e., $\dim \boldsymbol{x} = 2$.

**Theorem 3.** *Let* $n = 2$. *Then the vector field* $\boldsymbol{u}_0$ *is degenerate iff there exist a function* $\varphi_0 \colon \mathbb{R} \to \mathbb{R}$ *and real numbers* $b_1$, $b_2$, $c_1$, *and* $c_2$ *such that*

$$\begin{aligned} \{\boldsymbol{f}(\boldsymbol{u}_0(\boldsymbol{x}))\}_1 &= \phantom{-}b_2 \varphi_0(b_1 x_1 + b_2 x_2) + c_1, \\ \{\boldsymbol{f}(\boldsymbol{u}_0(\boldsymbol{x}))\}_2 &= -b_1 \varphi_0(b_1 x_1 + b_2 x_2) + c_2. \end{aligned} \tag{3.3}$$

*Proof.* The sufficiency is trivial. Indeed, if the vector field $\boldsymbol{f}(\boldsymbol{u}_0(\boldsymbol{x}))$ has the form (3.3), then the Jacobi matrix

$$\begin{pmatrix} \partial f_1(\boldsymbol{u}_0)/\partial x_1 & \partial f_1(\boldsymbol{u}_0)/\partial x_2 \\ \partial f_2(\boldsymbol{u}_0)/\partial x_1 & \partial f_2(\boldsymbol{u}_0)/\partial x_2 \end{pmatrix} = \begin{pmatrix} b_1 b_2 \varphi_0' & b_2^2 \varphi_0' \\ -b_1^2 \varphi_0' & -b_2 b_1 \varphi_0' \end{pmatrix}$$

is nilpotent.

Necessity. Let $c_1$ and $c_2$ be the mean values of $f_1(\boldsymbol{u}_0(\boldsymbol{x}))$ and $f_2(\boldsymbol{u}_0(\boldsymbol{x}))$, respectively. Since $\operatorname{div} \boldsymbol{f}(\boldsymbol{u}_0(\boldsymbol{x})) = 0$, there exists a function $\psi : \mathbb{R}^2 \to \mathbb{R}$ such that $\operatorname{rot} \psi = \boldsymbol{f}(\boldsymbol{u}_0(\boldsymbol{x})) - \binom{c_1}{c_2}$, where $\operatorname{rot} \psi = \binom{\partial\psi/\partial x_2}{-\partial\psi/\partial x_1}$. We note that the function $\psi(x_1, x_2)$ is $\mathbb{T}^2$-periodic, and hence bounded. Since the determinant of the Jacobi matrix of $\boldsymbol{f}(\boldsymbol{u}_0(\boldsymbol{x}))$ is zero, we have that determinant of the Hessian of $\psi$ is zero. Consider the graph of the function $\psi$ in $\mathbb{R}^3$. The Gaussian curvature of this surface is given by the formula (see [3], I, chpt. 2)

$$K = \frac{\psi_{xx}\psi_{yy} - \psi_{xy}^2}{(1 + \psi_x^2 + \psi_y^2)^2}.$$

Therefore, $K = 0$. Now we use the fact that any complete surface of constant zero Gaussian curvature is a cylinder over a flat curve (see [13]; [14], chpt. 5). Since the function $\psi$ is bounded, every generator of this cylinder is a horizontal line, hence it's equation can be written in the form

$$\begin{cases} b_1 x_1 + b_2 x_2 = const, \\ z = \tilde{\psi}(const). \end{cases}$$

We conclude that

$$\psi(x_1, x_2) = \tilde{\psi}(b_1 x_1 + b_2 x_2)$$

and (3.3) follows with $\varphi_0 = \tilde{\psi}'$. □

**Corollary 2.** *Suppose, $m = n = 2$, $\boldsymbol{f}(\boldsymbol{u}) \equiv \boldsymbol{u}$, and $\boldsymbol{h} \equiv 0$; then the solution of the Cauchy problem* (1.1), (3.3) *remains of the form* (3.3)*:*

$$\boldsymbol{u}(t, \boldsymbol{x}) = \begin{pmatrix} b_2 \\ -b_1 \end{pmatrix} \varphi(t, b_1 x_1 + b_2 x_2) + \begin{pmatrix} c_1 \\ c_2 \end{pmatrix}$$

*where the function $\varphi$ satisfies the equation*

$$\varphi_t + (b_1 c_1 - b_2 c_2)\varphi' = (b_1^2 + b_2^2)\nu\varphi''.$$

In this case we have $\nu$-independent upper bounds for derivatives of the solution.

Further on we shall use the polynomial

$$P_{\boldsymbol{x}}(t) = t^n \chi_{\boldsymbol{x}}\left(\tfrac{-1}{t}\right) = \det\left(\delta_{ij} + \tfrac{\partial \tilde{f}_i}{\partial x_j} t\right) = 1 + I_1(\boldsymbol{x})t + I_2(\boldsymbol{x})t^2 + \cdots + I_n(\boldsymbol{x})t^n, \quad (3.4)$$

rather than a characteristic polynomial.

### 3.2. General idea

In this subsection we present an auxiliary theorem from which we then derive Theorem 2. This auxiliary theorem is technically complicated. Here we deal mainly with general ideas, and postpone the technicalities to the next subsection.

We denote the right hand side of (2.1) by $g_i$:

$$\frac{\partial}{\partial t} u_i + \sum_{j=1}^{n} f_j(u_1, \ldots, u_m) \frac{\partial u_i}{\partial x_j} = g_i(t, x_1, \ldots, x_n). \quad (3.5)$$

**Theorem 4.** *Let $\boldsymbol{f}\colon \mathbb{R}^m \to \mathbb{R}^n$ be a $C^1$-smooth map and let $\boldsymbol{u}_0\colon \mathbb{T}^n \to \mathbb{R}^m$ be a $C^1$-smooth vector field.*

**1)** *If $\boldsymbol{u}_0$ is non-degenerate, then there exist $T = T(\boldsymbol{f}, \boldsymbol{u}_0) < \infty$ and $c = c(\boldsymbol{f}, \boldsymbol{u}_0) > 0$ such that for any $C^1$-smooth vector field $\boldsymbol{u}\colon [0,T] \times \mathbb{T}^n \to \mathbb{R}^m$ with $\boldsymbol{u}(0,\boldsymbol{x}) = \boldsymbol{u}_0(\boldsymbol{x})$ we have*

$$\int_0^T \sup_{\boldsymbol{x}\in\mathbb{T}^n} |\boldsymbol{g}(\tau,\boldsymbol{x})| d\tau \geqslant c\,, \tag{3.6}$$

*where $\boldsymbol{g}$ is given by (3.5).*

**2)** *If $\boldsymbol{u}_0$ is degenerate, then there is a $C^1$-smooth vector field $\boldsymbol{u}\colon [0,+\infty) \times \mathbb{T}^n \to \mathbb{R}^m$ such that $\boldsymbol{u}(0,\boldsymbol{x}) = \boldsymbol{u}_0(\boldsymbol{x})$ and $g_i(t,\boldsymbol{x}) \equiv 0$.*

*Proof.* **1)** Without loss of generality it can be assumed that $\boldsymbol{u}(t,\boldsymbol{x})$ is defined for all $t \geqslant 0$. Consider the flow on the cylinder $\mathbb{T}^n \times [0,\infty)$ generated by the vector field $\boldsymbol{f}(\boldsymbol{u})$. In other words we consider the Cauchy problem

$$\tfrac{d}{dt}\gamma(t,\boldsymbol{\xi}) = \boldsymbol{f}(\boldsymbol{u}(t,\boldsymbol{\xi}))$$

with the initial state $\gamma(0,\boldsymbol{\xi}) = \boldsymbol{\xi}$. Here $\boldsymbol{\xi}$ is the Lagrange coordinate of the flow $\gamma$.

For any fixed time $t$ we have a map $\gamma(t,\cdot) : \mathbb{T}^n \to \mathbb{T}^n$. Since $\gamma(t,\cdot)$ is a continuous family of diffeomorphisms, equal identity for $t = 0$, then its Jacobian is everywhere positive.

Combining the chain rule and (3.5), we obtain

$$\tfrac{d}{dt}\boldsymbol{u}(t,\gamma(t,\boldsymbol{\xi})) = \boldsymbol{g}(t,\gamma(t,\boldsymbol{\xi}))\,.$$

Suppose $\boldsymbol{g}(\cdot) \equiv 0$; then $\boldsymbol{f}\big(\boldsymbol{u}(t,\gamma(t,\boldsymbol{\xi}))\big) \equiv \boldsymbol{f}\big(\boldsymbol{u}(0,\gamma(0,\boldsymbol{\xi}))\big)$ and $\gamma(t,\boldsymbol{\xi}) = \gamma^{\mathbf{0}}(t,\boldsymbol{\xi})$, where

$$\gamma^{\mathbf{0}}(t,\boldsymbol{\xi}) = \boldsymbol{\xi} + t\boldsymbol{f}(\boldsymbol{u}_0(\boldsymbol{\xi}))\,. \tag{3.7}$$

It follows that if the function $\boldsymbol{g}$ is small, then the flow $\gamma(t,\boldsymbol{\xi})$ is close (in the $C^0$-norm) to the map (3.7). For a detailed proof of this fact, we refer to the next subsection. For the time being we simply note that this is a consequence of the following inequality:

$$\big|\boldsymbol{u}(t,\gamma(t,\boldsymbol{\xi})) - \boldsymbol{u}(0,\gamma(0,\boldsymbol{\xi}))\big| \leqslant \int_0^t \sup_{\boldsymbol{x}\in\mathbb{T}^n} |\boldsymbol{g}(\tau,\boldsymbol{x})| d\tau\,.$$

Since for each $k = 1,\dots,n$, we have $\int_{\mathbb{T}^n} I_k(\boldsymbol{x})d\boldsymbol{x} = 0$ (see lemma 3) and since some of the $I_k$ are not identically zero (due to the non-degeneracy of $\boldsymbol{u}_0$), we obtain that there exists a point $\boldsymbol{x}^\star \in \mathbb{T}^n$ and a number $l \in [1,\dots,n]$ such that $I_l(\boldsymbol{x}^\star) < 0$ and $I_k(\boldsymbol{x}^\star) = 0$ for $k > l$.

The Jacobian of (3.7) is expressed by polynomial (3.4). For the time $t = T$ at the point $\boldsymbol{\xi} = \boldsymbol{x}^\star$, we have

$$\det\Big(\tfrac{\partial\gamma^0}{\partial\boldsymbol{\xi}}\Big|_{\substack{t=T\\ \boldsymbol{\xi}=\boldsymbol{x}^\star}}\Big) = P_{\boldsymbol{x}^\star}(T) = 1 + I_1(\boldsymbol{x}^\star)T + I_2(\boldsymbol{x}^\star)T^2 + \cdots + I_l(\boldsymbol{x}^\star)T^l\,.$$

We take large enough time $T$ such that this Jacobian is negative. Suppose that

$$\int_0^T \sup_{\boldsymbol{x}\in\mathbb{T}^n} |\boldsymbol{g}(\tau,\boldsymbol{x})|d\tau < c,$$

where $c$ is a small enough number; then the map $\gamma(T,\cdot)$ is close to the map $\gamma^0(T,\cdot)$ with a negative Jacobian at the point $\boldsymbol{x}^\star$. Taking $c$ small enough we have a contradiction with the positivity of the Jacobian of the map $\gamma(T,\cdot)$. (See the next part of this section for more details).

**2)** Consider the map (3.7). For any fixed $t$ we have a ($C^1$-smooth) map $\boldsymbol{\gamma}^{\mathbf{0}}(t,\cdot)\colon \mathbb{T}^n \to \mathbb{T}^n$. We note that this map is a ($C^1$-smooth) diffeomorphism of the torus $\mathbb{T}^n$ iff the Jacobian of $\boldsymbol{\gamma}^{\mathbf{0}}(t,\cdot)$ is everywhere positive. Indeed, if this map is a diffeomorphism, then the Jacobian is not vanishing, hence it has the same sign for all points, and this sign is positive since the map is homotopic to the identity map. If the Jacobian is everywhere positive then by the inverse function theorem we have that $\boldsymbol{\gamma}^{\mathbf{0}}(t,\cdot)$ is a (local) diffeomorphism in a neighbourhood of any point. Since the Jacobian is everywhere positive, the number of preimages of any point $\boldsymbol{z}$ is finite and equals the degree of the map. (see [3], II, chpt. 3). On the other hand, the degree of the map $\boldsymbol{\gamma}^{\mathbf{0}}(t,\cdot)$ is equal to 1, since this map is homotopic to the identical map $\boldsymbol{\gamma}^{\mathbf{0}}(0,\cdot)$. (see [3], II, chpt. 3). Hence each point $\boldsymbol{z}$ has a unique pre-image $(\boldsymbol{\gamma}^{\mathbf{0}})^{-1}(t,\boldsymbol{z})$.

The Jacobian of the map $\boldsymbol{\gamma}^{\mathbf{0}}(t,\cdot)$ at a point $\boldsymbol{x}$ is expressed by the polynomial (3.4):

$$\left(\tfrac{\partial\gamma^0}{\partial\boldsymbol{\xi}}\right) = P_{\boldsymbol{x}}(t).$$

If $\boldsymbol{u}_0$ is degenerate then $P_{\boldsymbol{x}}(t) \equiv 1$. Hence the vector field

$$\boldsymbol{u}(t,\boldsymbol{x}) = \boldsymbol{u}_0\big((\boldsymbol{\gamma}^{\mathbf{0}})^{-1}(t,\boldsymbol{x})\big)$$

is well defined and satisfies $\partial_t\boldsymbol{u} + \nabla_{\boldsymbol{f}(\boldsymbol{u})}\boldsymbol{u} = 0$ as the second part of theorem 4 states. □

Let us turn to the proof of Theorem 2. Let $\boldsymbol{u}$ satisfy equation (1.1). Suppose that $H(t) < \frac{c}{2}$, where $H(t)$ is defined by (1.3); then we obtain

$$\int_0^T \sup_{\boldsymbol{x}\in\mathbb{T}^n} |\nu\Delta\boldsymbol{u}(\tau,\boldsymbol{x})|\, d\tau > \frac{c}{2}\,.$$

Hence we have inequality (1.7) for $k=2$ with the constant $r_2 = \frac{c}{2nT}$. We now fix the index $j$ to the value for which the maximum in (1.7) for $k=2$ is achieved. For this index we have

$$\frac{1}{T}\int_0^T \sup_{\boldsymbol{x}\in\mathbb{T}^n} \left|\tfrac{\partial^2\boldsymbol{u}}{\partial x_j^2}(\tau,\boldsymbol{x})\right|d\tau \geqslant \tfrac{c}{2nT}\,\tfrac{1}{\nu}\,. \tag{3.8}$$

To complete the proof we need the interpolation inequality

$$\left|\tfrac{\partial^2\boldsymbol{u}}{\partial x_j^2}(\tau,\cdot)\right|_{L_\infty} \leqslant C_{k,2}\,|\boldsymbol{u}(\tau,\cdot)|_{L_\infty}^{\frac{k-2}{k}}\,\left|\tfrac{\partial^k\boldsymbol{u}}{\partial x_j^k}(\tau,\cdot)\right|_{L_\infty}^{\frac{2}{k}}. \tag{3.9}$$

For its proof with the best possible constants for the $1D$ case ($m = n = 1$) see [10]. The case of arbitrary dimensions can be reduced to the $1D$ case by considering the function $v(x_j) = \sum_{i=1}^m a_i u_i(x_1^0, \ldots, x_{j-1}^0, x_j, x_{j+1}^0, \ldots, x_n^0)$, where $\boldsymbol{x^0} = \boldsymbol{x^0}(\tau) \in \mathbb{T}^n$ is the maximum point of the $\left|\frac{\partial^2 \boldsymbol{u}}{\partial x_j^2}(\tau, \boldsymbol{x})\right|$ and $\boldsymbol{a}$ is a constant unit vector in $\mathbb{R}^m$, proportional to $\frac{\partial^2 \boldsymbol{u}}{\partial x_j^2}(\tau, \boldsymbol{x^0})$. We note that this reduction preserves the Kolmogorov's constants.

Using (3.9) and inequality (2.2), we obtain

$$\left|\frac{\partial^k \boldsymbol{u}}{\partial x_j^k}(\tau, \cdot)\right|_{L_\infty} \geqslant \frac{\left|\frac{\partial^2 \boldsymbol{u}}{\partial x_j^2}(\tau, \cdot)\right|_{L_\infty}^{k/2}}{\sqrt{3}\big(|\boldsymbol{u}_0| + H(\tau)\big)^{\frac{k-2}{2}}}. \tag{3.10}$$

Here we have used the inequality $(C_{k,2})^{k/2} \leqslant \sqrt{3}$, which can be easily proved using Kolmogorov's explicit representation (see [10]) via the Hölder inequality. Integrating (3.10) we obtain

$$\begin{aligned}\frac{1}{T}\int_0^T \left|\frac{\partial^k \boldsymbol{u}}{\partial x_j^k}(\tau, \cdot)\right|_{L_\infty} d\tau &\geqslant \frac{\frac{1}{T}\int_0^T \left|\frac{\partial^2 \boldsymbol{u}}{\partial x_j^2}(\tau, \cdot)\right|_{L_\infty}^{k/2} d\tau}{\sqrt{3}\big(|\boldsymbol{u}_0| + H(T)\big)^{\frac{k-2}{2}}} \\ &\geqslant \frac{\left(\frac{1}{T}\int_0^T \left|\frac{\partial^2 \boldsymbol{u}}{\partial x_j^2}(\tau, \cdot)\right|_{L_\infty} d\tau\right)^{k/2}}{\sqrt{3}\big(|\boldsymbol{u}_0| + H(T)\big)^{\frac{k-2}{2}}} \overset{(3.8)}{\geqslant} \frac{\left(\frac{c}{2nT\nu}\right)^{k/2}}{\sqrt{3}\big(|\boldsymbol{u}_0| + H(T)\big)^{\frac{k-2}{2}}}.\end{aligned}$$

This concludes the proof of (1.7) for $k > 2$ with the constants

$$r_k = \frac{\left(\frac{c}{2nT}\right)^{\frac{k}{2}}}{\sqrt{3}\big(|\boldsymbol{u}_0| + H(T)\big)^{\frac{k-2}{2}}}. \tag{3.11}$$

In the next subsection we will specify the values of $T$ and $c$ (see (3.16) and (3.33) respectively).

### 3.3. Technicalities

In this subsection we introduce a more general approach to the estimates of Theorem 2 which applies to the non-periodic case. From the previous subsection we already know that the crucial condition for Theorem 2 is the "negativity" rather than the non-degeneracy of the matrix $\frac{\partial \boldsymbol{f}(\boldsymbol{u}_0(\boldsymbol{x}))}{\partial \boldsymbol{x}}$. Let $\boldsymbol{u} : [0, \infty) \times \mathbb{R}^n \to \mathbb{R}^m$ be a $C^1$-smooth vector-valued function and $\boldsymbol{f} : \mathbb{R}^m \to \mathbb{R}^n$ be a $C^1$-smooth map. In this subsection $\boldsymbol{x} = (x_1, \ldots, x_n)$ are coordinates in $\mathbb{R}^n$.

We define $g_i : [0, \infty) \times \mathbb{R}^n \to \mathbb{R}$, $(i = 1, \ldots, m)$ to satisfy:

$$\partial_t u_i + \sum_{j=1}^n f_j(u_1, \ldots, u_m)\frac{\partial u_i}{\partial x_j} = g_i(t, x_1, \ldots, x_n). \tag{3.12}$$

Let $\tilde{\boldsymbol{f}}(\boldsymbol{x}) = \boldsymbol{f}(\boldsymbol{u}(0,\boldsymbol{x}))$. If this function is $C^2$-smooth then we consider the norm

$$|\tilde{\boldsymbol{f}}|_2^2 = \sup_{\boldsymbol{x}} \sum_{i,j,k=1}^{n} \left| \frac{\partial^2 \tilde{f}_i}{\partial x_j \partial x_k} \right|^2 . \tag{3.13}$$

For any $\boldsymbol{u} \in \mathbb{R}^m$ the derivative $\boldsymbol{\nabla_u f} : \mathbb{R}^m \to \mathbb{R}^n$ is a linear map. For any domain $E \subset \mathbb{R}^m$ we denote

$$\|\boldsymbol{\nabla_u f}\|_E = \sup_{\boldsymbol{u}\in E} \|\boldsymbol{\nabla_u f}\| = \sup_{\boldsymbol{u}\in E} \max_{|\boldsymbol{v}|=1} \big|(\boldsymbol{\nabla_u f})\boldsymbol{v}\big|, \tag{3.14}$$

where $\{(\boldsymbol{\nabla_u f})\boldsymbol{v}\}_j = \sum_{i=1\dots m} \frac{\partial f_j}{\partial u_i} v_i$. We note that for the Burgers' (=NS') nonlinearity (i.e., $m = n$ and $\boldsymbol{f} \equiv \boldsymbol{u}$) we have $\|\boldsymbol{\nabla_u f}\| \equiv 1$.

**Theorem 5.** *Suppose that $\boldsymbol{u} : [0,\infty) \times \mathbb{R}^n \to \mathbb{R}^m$ and $\boldsymbol{f} : \mathbb{R}^m \to \mathbb{R}^n$ are $C^1$-smooth. Let $\varepsilon$ be a positive real number and $l \in \{1,2,\dots,n\}$. Suppose that there exists $\boldsymbol{x}^\star \in \mathbb{R}^n$ such that $I_l(\boldsymbol{x}^\star) = -\varepsilon < 0$ and that for $k = l+1,\dots,n$ we have $I_k(\boldsymbol{x}^\star) = 0$, where $I_i$ are defined in (3.1). Let*

$$|\tilde{\boldsymbol{f}}|_1 = \max_{i,j} \left| \frac{\partial \tilde{f}_i}{\partial x_j}(\boldsymbol{x}^\star) \right| , \tag{3.15}$$

*where $\tilde{\boldsymbol{f}}(\boldsymbol{x}) = \boldsymbol{f}(\boldsymbol{u}(0,\boldsymbol{x}))$ and let*

$$T = 2^n l^{l/2} \frac{|\tilde{\boldsymbol{f}}|_1^{l-1}}{\varepsilon} . \tag{3.16}$$

*Then there exists a positive function $c_2(c_1)$ such that if*

$$\int_0^T \sup_{\boldsymbol{x}\in\mathbb{R}^n} |\boldsymbol{g}(\tau,\boldsymbol{x})|\, d\tau < c_1 , \tag{3.17}$$

*then*

$$\int_0^T (T-\tau) \sup_{\boldsymbol{x}\in\mathbb{R}^n} |\boldsymbol{g}(\tau,\boldsymbol{x})|\, d\tau \geqslant c_2(c_1) . \tag{3.18}$$

*If the function $\tilde{\boldsymbol{f}}$ is $C^2$-smooth and the norm (3.13) is finite, then one can take*

$$c_2(c_1) = \frac{\frac{1}{4} l^l}{n^{2n-2}(|\tilde{\boldsymbol{f}}|_1 T)^{2n-2l}\, \|\boldsymbol{\nabla_u f}\|_{B(|\boldsymbol{u}_0|+c_1)} |\tilde{\boldsymbol{f}}|_2 T} . \tag{3.19}$$

*Here $B(r)$ denotes the ball in $\mathbb{R}^m$ of radius $r$ in the Euclidean norm, centered at the origin.*

*Proof.* First of all we note that $\boldsymbol{f}(\boldsymbol{u}(t,\boldsymbol{x}))$ and $\frac{\partial}{\partial \boldsymbol{x}}\boldsymbol{f}(\boldsymbol{u}(t,\boldsymbol{x}))$ are a continuous vector function and a continuous matrix function, respectively. It follows from this that

1° ∃! solution of the Cauchy problem for the following ODE in $R^n$:

$$\tfrac{d}{dt}\gamma(t,\boldsymbol{\xi}) = \boldsymbol{f}(\boldsymbol{u}(t,\boldsymbol{\xi})) , \tag{3.20}$$

with the initial state $\gamma(0,\boldsymbol{\xi}) = \boldsymbol{\xi} \in \mathbb{R}^n$.

2° This solution $\gamma \in C^1\big([0,T] \times \mathbb{R}^n; \mathbb{R}^n\big)$.

3° This solution satisfies $det\left(\frac{\partial\boldsymbol{\gamma}(t,\boldsymbol{\xi})}{\partial\boldsymbol{\xi}}\right) = \exp\int_0^t\sum_{i=0}^n\frac{\partial}{\partial\gamma_i}f_i\big(\boldsymbol{u}(t,\boldsymbol{\gamma}(t,\boldsymbol{\xi}))\big)$, see [8].

The positiveness of the Jacobian in 3° implies inequality (3.18). The rest of this subsection is devoted to proving this fact.

In the proof of Proposition 1 below we will use the fact that there exists a continuous second derivative $\frac{d^2}{dt^2}\boldsymbol{\gamma}(t,\boldsymbol{\xi})$. This follows from the existence and continuity of the first partial derivative $\frac{\partial}{\partial t}\boldsymbol{f}(\boldsymbol{u}(t,\boldsymbol{x}))$.

For the quantities (3.1) we have:

$$|I_k| \leqslant \tbinom{n}{k}k^{k/2}|\tilde{\boldsymbol{f}}|_1^k\,. \tag{3.21}$$

Indeed, the right hand side of (3.1) contains $\binom{n}{k}$ terms and each of them is no greater than $k^{k/2}|\tilde{\boldsymbol{f}}|_1^k$. Here we have used the fact that the volume of a $k$-dimensional parallelepiped with sides of length less than or equal to $\sqrt{k}|\tilde{\boldsymbol{f}}|_1$ is no greater than $(\sqrt{k}|\tilde{\boldsymbol{f}}|_1)^k$.

In the proofs of Propositions 2 and 3 below we will use the inequality

$$T|\tilde{\boldsymbol{f}}|_1 > 1\,. \tag{3.22}$$

It follows from (3.16) and (3.21) with $k = l$ since $\varepsilon = |I_l(\boldsymbol{x}^\star)|$.

We fix $T$ at the value given in (3.16). For any $t$, $\boldsymbol{\gamma}(t,\boldsymbol{x})$ defines a mapping from $\mathbb{R}^n$ into itself. We take $t = T$ and decompose this mapping as follows:

$$\gamma_i(T,\boldsymbol{x}) = p_i(\boldsymbol{x}) + q_i(\boldsymbol{x})\,, \qquad i = 1,\ldots,n. \tag{3.23}$$

Here $\boldsymbol{p}(\boldsymbol{x})$ comprises the zeroth and the first terms of the Tailor expansion:

$$p_i(\boldsymbol{x}) = \gamma_i(0,\boldsymbol{x}) + \frac{d}{dt}\gamma_i(0,\boldsymbol{x})T = \boldsymbol{x} + \tilde{\boldsymbol{f}}(\boldsymbol{x})T\,. \tag{3.24}$$

The remainder of the Tailor expansion can be represented as

$$q_i(\boldsymbol{x}) = \int_0^T (T-\tau)\frac{d^2}{dt^2}\gamma_i(\tau,\boldsymbol{x})\,d\tau\,. \tag{3.25}$$

**Proposition 1.** *For the Euclidean norm of the vector $\boldsymbol{q}$ we have:*

$$|\boldsymbol{q}| \leqslant \|\boldsymbol{\nabla_u f}\|_{B(|\boldsymbol{u}_0|+c_1)}\int_0^T (T-\tau)\sup_{\boldsymbol{x}\in\mathbb{R}^n}|\boldsymbol{g}(\tau,\boldsymbol{x})|\,d\tau\,. \tag{3.26}$$

We recall that $|\cdot|$ denotes the Euclidean norm.

*Proof.* Combining the chain rule with (3.20) and (3.12) we have

$$\frac{d}{dt}u_i(t,\boldsymbol{\gamma}(t,\boldsymbol{x})) = g_i(t,\boldsymbol{\gamma}(t,\boldsymbol{x}))\,.$$

Using this equality and assumption (3.17) we obtain that the function $\boldsymbol{u}$ takes values in the ball $B(|\boldsymbol{u}_0| + c_1)$ if $t \leqslant T$. We calculate the second derivative of $\gamma_i$

(for $i = 1, \dots, n$):

$$\frac{d^2}{dt^2}\gamma_i(\tau,\boldsymbol{x}) = \frac{d}{dt} f_i(\boldsymbol{u}(t,\boldsymbol{\gamma}(t,\boldsymbol{x})))\Big|_{t=\tau} = \sum_{j=1}^{m} \frac{\partial f_i}{\partial u_j} g_j(\tau,\boldsymbol{\gamma}(\tau,\boldsymbol{x})) = \{(\nabla_{\boldsymbol{u}}\boldsymbol{f})\boldsymbol{g}\}_i \,.$$

From this formula we obtain

$$\left|\tfrac{d^2}{dt^2}\boldsymbol{\gamma}(\tau,\boldsymbol{x})\right| \leqslant \|\nabla_{\boldsymbol{u}}\boldsymbol{f}\|_{B(|\boldsymbol{u}_0|+c_1)} \sup_{\boldsymbol{x}\in\mathbb{R}^n} |\boldsymbol{g}(\tau,\boldsymbol{x})|.$$

Multiplying this inequality by $T-\tau$ and using (3.25) we arrive at (3.26). □

Consider the linearization of the map $\boldsymbol{p}$ at the point $\boldsymbol{x}^\star$ (we recall that $\boldsymbol{x}^\star$ is the point where the leading non zero $I_k$ is negative):

$$p_i(\boldsymbol{x}) = p_i(\boldsymbol{x}^\star) + \sum_{j=1}^{n} \frac{\partial p_i}{\partial x_j}(\boldsymbol{x}^\star)\vartriangle x_j + \phi_i(\vartriangle\boldsymbol{x})\,, \tag{3.27}$$

where $\vartriangle\boldsymbol{x} = \boldsymbol{x} - \boldsymbol{x}^\star$ and $\boldsymbol{\phi}(\vartriangle\boldsymbol{x}) = o(\vartriangle\boldsymbol{x})$.

We need to investigate the matrix

$$A = \tfrac{\partial p_i}{\partial x_j}\Big|_{\boldsymbol{x}=\boldsymbol{x}^\star} = \delta_{ij} + T\tfrac{\partial \tilde f_i}{\partial x_j}\Big|_{\boldsymbol{x}=\boldsymbol{x}^\star}, \tag{3.28}$$

which is the linear part of the right hand side of (3.27).

**Proposition 2.** *The determinant of the matrix A is negative and bounded away from zero:*

$$\det A \leqslant -2^{n-1} l^{l/2} (T|\tilde{\boldsymbol{f}}|_1)^{l-1}. \tag{3.29}$$

*Proof.* The determinant is expressed by polynomial (3.4): $det\, A = P_{\boldsymbol{x}^\star}(T)$. Using (3.21) we have

$$\begin{aligned}\det A &\leqslant 1 + \tbinom{n}{1} 1^{1/2} T|\tilde{\boldsymbol{f}}|_1 + \dots + \tbinom{n}{l-1}(l-1)^{(l-1)/2}(T|\tilde{\boldsymbol{f}}|_1)^{l-1} - \varepsilon T^l \\ &\overset{(*)}{\leqslant} \tfrac{1}{2} 2^n l^{l/2} (T|\tilde{\boldsymbol{f}}|_1)^{l-1} - \varepsilon T^l = T^{l-1}\big(\tfrac{1}{2}2^n l^{l/2} |\tilde{\boldsymbol{f}}|_1^{l-1} - \varepsilon T\big).\end{aligned}$$

Using (3.16) we arrive at (3.29). It remains to explain inequality $(*)$. For $l = 1$ it follows from the trivial inequality $1 \leqslant 2^{n-1}$. For $l \geqslant 2$ we use the simple fact $(l-1)^{(l-1)/2} \leqslant \frac{1}{2} l^{l/2}$ and inequality (3.22) to get

$$\begin{aligned}&1 + \tbinom{n}{1} 1^{1/2} T|\tilde{\boldsymbol{f}}|_1 + \dots + \tbinom{n}{l-1}(l-1)^{(l-1)/2}(T|\tilde{\boldsymbol{f}}|_1)^{l-1} \\ &\quad \leqslant \Big(\tbinom{n}{0} + \tbinom{n}{1} + \dots + \tbinom{n}{l-1}\Big)(l-1)^{(l-1)/2}(T|\tilde{\boldsymbol{f}}|_1)^{l-1} \leqslant 2^n \tfrac{1}{2} l^{l/2} (T|\tilde{\boldsymbol{f}}|_1)^{l-1}.\end{aligned}$$

□

**Proposition 3.** *With the matrix norm* $\|\cdot\|$ *we have*

$$\|A^{-1}\|^{-1} \geqslant \frac{l^{l/2}}{n^{n-1}(T|\tilde{\boldsymbol{f}}|_1)^{n-l}}\,. \tag{3.30}$$

*Proof.* Since the numbers $\|A^{-1}\|^{-2}$ and $\|A\|^2$ are, respectively, the minimal and maximal eigenvalues of the matrix $A^tA$, we have $\|A^{-1}\|^{-1} \geqslant |\det A| \, \|A\|^{1-n}$. Using the inequalities $\|A\| \leqslant n \max_{i,j} |A_{ij}|$, $|A_{ij}| \leqslant 2T|\tilde{\boldsymbol{f}}|_1$ (the second inequality follows by (3.28) and (3.22)) and (3.29) we arrive at (3.30). □

Since $\phi_i(\Delta \boldsymbol{x}) = o(\Delta \boldsymbol{x})$, where $\phi_i(\Delta \boldsymbol{x})$ is the remainder term in (3.27), there exists $r > 0$ such that

$$|\phi_i(\Delta \boldsymbol{x})| \leqslant \tfrac{1}{2}\|A^{-1}\|^{-1} r \quad \text{for } |\Delta \boldsymbol{x}| \leqslant r. \tag{3.31}$$

Consider the sphere $S_r(\boldsymbol{x}^\star)$ with the centre at the point $\boldsymbol{x}^\star$ and with the radius $r$.

**Proposition 4.** *There exists $\boldsymbol{x_0} \in S_r(\boldsymbol{x}^\star)$ such that $|q(\boldsymbol{x_0})| \geqslant \frac{1}{2}\|A^{-1}\|^{-1} r$.*

*Proof.* Suppose that $|\boldsymbol{q}| < \frac{1}{2}\|A^{-1}\|^{-1} r$ on $S_r(\boldsymbol{x}^\star)$. Let $\boldsymbol{\rho}(\Delta \boldsymbol{x}) = \boldsymbol{\phi}(\Delta \boldsymbol{x}) + \boldsymbol{q}(\boldsymbol{x}^\star + \Delta \boldsymbol{x})$. Then due to (3.23) and (3.27) we obtain:

$$\boldsymbol{\gamma}(T, \boldsymbol{x}^\star + \Delta \boldsymbol{x}) = \boldsymbol{p}(\boldsymbol{x}^\star) + A\Delta \boldsymbol{x} + \boldsymbol{\rho}(\Delta \boldsymbol{x}) \,.$$

We recall that $\boldsymbol{x}^\star + \Delta \boldsymbol{x} = \boldsymbol{x}$. Using the inequality $|\boldsymbol{\rho}(\Delta \boldsymbol{x})| < \|A^{-1}\|^{-1} r$ and (3.30), we have

$$|A\Delta \boldsymbol{x}| \geqslant \|A^{-1}\|^{-1} |\Delta \boldsymbol{x}| > |\boldsymbol{\rho}(\Delta \boldsymbol{x})| \qquad \text{for } |\Delta \boldsymbol{x}| = r \text{ , i.e., } \boldsymbol{x} \in S_r(\boldsymbol{x}^\star)\,.$$

From this inequality it follows that the Gauss spherical map $\Gamma : S_r(\boldsymbol{x}^\star) \to S_1(\boldsymbol{0})$

$$\Delta \boldsymbol{x} \mapsto \frac{A\Delta \boldsymbol{x} + \boldsymbol{\rho}(\Delta \boldsymbol{x})}{|A\Delta \boldsymbol{x} + \boldsymbol{\rho}(\Delta \boldsymbol{x})|} \tag{3.32}$$

is well defined and is homotopic to the map

$$\Delta \boldsymbol{x} \mapsto \frac{A\Delta \boldsymbol{x}}{|A\Delta \boldsymbol{x}|}\,.$$

Hence the degrees of these maps coincide and are equal to sign det $A = -1$. (see [3], II, chpt. 3). On the other hand, the map (3.32) can be written as

$$\Delta \boldsymbol{x} \mapsto \frac{\boldsymbol{\gamma}(\boldsymbol{x}^\star + \Delta \boldsymbol{x}) - \boldsymbol{p}(\boldsymbol{x}^\star)}{|\boldsymbol{\gamma}(\boldsymbol{x}^\star + \Delta \boldsymbol{x}) - \boldsymbol{p}(\boldsymbol{x}^\star)|}\,.$$

Since the Jacobian of $\boldsymbol{\gamma}$ does not vanish, then the degree of this map is equal to

$$\sum_{\substack{\boldsymbol{y} \in \boldsymbol{\gamma}^{-1}(\boldsymbol{p}(\boldsymbol{x}^\star)) \\ \boldsymbol{y} \in B(r) + \boldsymbol{x}^\star}} \text{sign det} \left( \frac{\partial \boldsymbol{\gamma}(T, \boldsymbol{\xi})}{\partial \boldsymbol{\xi}} \Big|_{\boldsymbol{\xi} = \boldsymbol{y}} \right)$$

(see [3], II, Chpt. 3). This number is nonnegative, so we got a contradiction which proves Proposition 4. □

Using (3.26) we arrive at (3.18) with $c_2(c_1) = \dfrac{\|A^{-1}\|^{-1} r}{2\,|\!|\!|\nabla_{\boldsymbol{u}} \boldsymbol{f}|\!|\!|_{B(|\boldsymbol{u}_0| + c_1)}}$.

Suppose that the function $\tilde{\boldsymbol{f}}$ is $C^2$-smooth and the norm (3.13) is finite. Then the remainder term $\phi_i(\Delta \boldsymbol{x})$ can be written as

$$\phi_i(\Delta \boldsymbol{x}) = \int_0^1 (1-\theta) \sum_{j,k=1}^{n} \frac{\partial^2 p_i}{\partial x_j \partial x_k}(\boldsymbol{x}^\star + \theta \Delta \boldsymbol{x}) \Delta x_j \Delta x_k d\theta.$$

Hence

$$\sum_{i=1}^{n} \phi_i^2(\Delta \boldsymbol{x}) = \sum_{i=1}^{n} \Big( \sum_{j,k=1}^{n} \int_0^1 (1-\theta) \frac{\partial^2 p_i}{\partial x_j \partial x_k}(\boldsymbol{x}^\star + \theta \Delta \boldsymbol{x}) d\theta \Delta x_j \Delta x_k \Big)^2$$

$$\leqslant \int_0^1 (1-\theta)^2 \sum_{i,j,k=1}^{n} \Big( \frac{\partial^2 p_i}{\partial x_j \partial x_k}(\boldsymbol{x}^\star + \theta \Delta \boldsymbol{x}) \Big)^2 d\theta \sum_{j,k=1}^{n} (\Delta x_j \Delta x_k)^2 \leqslant \tfrac{1}{3} |\tilde{\boldsymbol{f}}|_2^2 T^2 \, |\Delta \boldsymbol{x}|^4 .$$

Now we see that (3.31) holds with

$$r = \frac{\left\| A^{-1} \right\|^{-1}}{2 |\tilde{\boldsymbol{f}}|_2 T}.$$

and (3.19) follows. Theorem 5 is proved. □

Using the inequality

$$\int_0^T \sup |\boldsymbol{g}(\tau,\cdot)| \, d\tau \geqslant \frac{1}{T} \int_0^T (T-\tau) \sup |\boldsymbol{g}(\tau,\cdot)| \, d\tau$$

we obtain $\int_0^T \sup |\boldsymbol{g}(\tau,\cdot)| \, d\tau \geqslant c$ with

$$c = \sup_{c_1 > 0} \min \left\{ c_1, \tfrac{c_2(c_1)}{T} \right\}. \tag{3.33}$$

Let $\boldsymbol{u}(0,\boldsymbol{x})$ be periodic with compact fundamental periodic domain $\mathbb{T}$. It follows that $\int_{\mathbb{T}} I_k(\boldsymbol{x}) d\boldsymbol{x} = 0$ (see Lemma 3); so our theorem is applicable for the periodic case iff not all $I_k(\boldsymbol{x})$ are identically zero. In this case we can put $l = \max\{k \in 1,2,\ldots,n : I_k \not\equiv 0\}$ and $\varepsilon = -\min I_l(\boldsymbol{x})$.

Hence, we have proved Theorem 2 with $r_k$ as in (3.11), and $c$ as in (3.33).

## 4. Fourier coefficients

In this section we present some results concerning behaviour of the Fourier coefficients of solutions for equation (1.1) which follow from what we have proved in the previous sections. These results are consistent with the so-called Kolmogorov–Obukhov (K-O) spectral asymptotics.

The K-O spectral law concerns distribution of the Fourier coefficients $\widehat{\boldsymbol{v}}_{\boldsymbol{s}}(t) = \frac{1}{\ell^{n/2}} \int_{\mathbb{T}^n} \boldsymbol{v}(t,\boldsymbol{x}) e^{\frac{-i2\pi \boldsymbol{s}\boldsymbol{x}}{\ell}} d\boldsymbol{x}$ of a velocity field $\boldsymbol{v}(t,\boldsymbol{x})$ which describes turbulent motion of $3D$ fluid with small viscosity. Due to the law, there exist non-negative constants $\kappa_1 < \kappa_2$ and $\varkappa$ such that for $\left(\frac{1}{\nu}\right)^{\kappa_1} < |\boldsymbol{s}| < \left(\frac{1}{\nu}\right)^{\kappa_2}$ (the inertial range) we have $\langle |\widehat{\boldsymbol{v}}_{\boldsymbol{s}}|^2 \rangle \sim \left(\frac{1}{|\boldsymbol{s}|}\right)^{\varkappa+n-1}$, i.e., the energy supported by wave-numbers $\boldsymbol{v}_{\boldsymbol{s}}$ on the sphere $\{r - Const \leqslant |\boldsymbol{s}| \leqslant r + Const\}$ behaves as $\left(\frac{1}{r}\right)^{\varkappa}$. Here $\langle\cdot\rangle$ denotes averaging over time and over a band of wave vectors. For $|\boldsymbol{s}| > \left(\frac{1}{\nu}\right)^{\kappa_2}$ (the dissipation range) the quantities $|\widehat{\boldsymbol{v}}_{\boldsymbol{s}}|$ decay faster than any power of $\frac{1}{|\boldsymbol{s}|}$. The theory does not say much about the energy range $|\boldsymbol{s}| < \left(\frac{1}{\nu}\right)^{\kappa_1}$ (see [5], Chpts 5, 6). The quantity $\nu^{\kappa_2}$ is called the Kolmogorov dissipation scale and $\varkappa$ is called the exponent of the K–O law.

Here and subsequently, $|\boldsymbol{s}|$ stands for the Euclidean norm of an integer vector $\boldsymbol{s} \in \mathbb{Z}^n$.

The K–O law is an *heuristic* law which applies to motion of $3D$ fluid, i.e., to solutions of the $3D$ Navier–Stokes system. Below we prove some rigorous results for solutions of the generalised Burgers equation 1.1, which is a K–O *type* spectral law. Roughly, we show that for time-averaged squared Fourier coefficients of the solutions we have $\kappa_2 \in [\frac{1}{2}, 1]$ and $\varkappa > 1$. Under the additional assumption $\kappa_1 < \frac{1}{[\frac{n}{2}]+3}$ we obtain the upper bound for the exponent of the spectral law: $\varkappa \leqslant 2[\frac{n}{2}]+5$. Under the assumption $\kappa_1 < \frac{1}{2}$ we have $\varkappa \leqslant \varkappa(\kappa_1)$, but our estimate $\varkappa(p)$ blows up to infinity as $p \to \frac{1}{2}$.

In [1] it is shown that for the 1D case the Kolmogorov dissipation scale is equal to $\nu$ (i.e., $\kappa_2 = 1$) and the exponent of the spectral law is equal to 2.

### 4.1. General situation

In this subsection we prove general lemmas, which provide information on the Fourier coefficients, if we know upper and lower bounds for the Sobolev norms. First steps in this direction were made by Kuksin [12] and most of the ideas in this section actually guided by the [12]. However we are adopt here a slightly different presentation and present some alternative proofs.

**Lemma 4.** *Suppose that there exist real numbers $k \geqslant 0$, $p''(k)$, $c''_k > 0$ and a set $\Upsilon''_k \subset (0,1]$ such that for any $\nu \in \Upsilon''_k$ we have*

$$\sum_{\boldsymbol{s}\in\mathbb{Z}^n} |\boldsymbol{s}|^{2k} \hat{a}^2_{\boldsymbol{s}}(\nu) \leqslant c''_k \left(\tfrac{1}{\nu}\right)^{2p''(k)}. \tag{4.1}$$

*Then for any positive real number $y$ and any $\nu \in \Upsilon''_k$ we have*

$$\sum_{|\boldsymbol{s}|\geqslant y} \hat{a}^2_{\boldsymbol{s}}(\nu) \leqslant c''_k y^{-2k} \nu^{-2p''(k)}. \tag{4.2}$$

*Proof.* For any positive real $y$ we have

$$\sum_{|\boldsymbol{s}|\geqslant y} \hat{a}^2_{\boldsymbol{s}} \leqslant y^{-2k} \sum_{|\boldsymbol{s}|\geqslant y} |\boldsymbol{s}|^{2k} \hat{a}^2_{\boldsymbol{s}} \leqslant y^{-2k} \sum_{\boldsymbol{s}\in\mathbb{Z}^n} |\boldsymbol{s}|^{2k} \hat{a}^2_{\boldsymbol{s}}.$$

Using (4.1) we arrive at (4.2). □

As a corollary, taking $y = \lambda_1(\frac{1}{\nu})^z$, for any positive real numbers $z$ and $\lambda_1 \leqslant \lambda_2 \leqslant +\infty$ and any $\nu \in \Upsilon''_k$ we have

$$\sum_{\lambda_1(\frac{1}{\nu})^z \leqslant |s| \leqslant \lambda_2(\frac{1}{\nu})^z} \hat{a}^2_{\boldsymbol{s}}(\nu) \leqslant \lambda_1^{-2k} c''_k \nu^{2kz-2p''(k)}.$$

**Lemma 5.** *Suppose that there exist real numbers $0 \leqslant k_1 < k < k_2$, $p''(k_1)$, $p'(k)$, $p''(k_2)$, $c''_{k_1} > 0$, $c'_k > 0$, $c''_{k_2} > 0$ and sets $\Upsilon''_{k_1}, \Upsilon'_k, \Upsilon''_{k_2} \subset (0,1]$ such that*

$$\sum_{\boldsymbol{s} \in \mathbb{Z}^n} |\boldsymbol{s}|^{2k_i} \hat{a}^2_{\boldsymbol{s}}(\nu) \leqslant c''_{k_i} \left(\tfrac{1}{\nu}\right)^{2p''(k_i)} \quad \text{for any } \nu \in \Upsilon''_{k_i}, \quad i = 1,2 \tag{4.3}$$

*and*

$$\sum_{\boldsymbol{s} \in \mathbb{Z}^n} |\boldsymbol{s}|^{2k} \hat{a}^2_{\boldsymbol{s}}(\nu) \geqslant c'_k \left(\tfrac{1}{\nu}\right)^{2p'(k)} \quad \text{for any } \nu \in \Upsilon'_k. \tag{4.4}$$

*Then for any $\mu \in (0,1)$ and any real $A \leqslant \bar{A}(\nu)$ and $B \geqslant \bar{B}(\nu)$ where*

$$\bar{A}(\nu) = \left(\tfrac{\mu}{2}\tfrac{c'_k}{c''_{k_1}}\right)^{\frac{1}{2k-2k_1}} \left(\tfrac{1}{\nu}\right)^{\frac{p'(k)-p''(k_1)}{k-k_1}} \quad \text{and} \quad \bar{B}(\nu) = \left(\tfrac{2}{\mu}\tfrac{c''_{k_2}}{c'_k}\right)^{\frac{1}{2k_2-2k}} \left(\tfrac{1}{\nu}\right)^{\frac{p''(k_2)-p'(k)}{k_2-k}} \tag{4.5}$$

*and any $\nu \in \Upsilon''_{k_1} \cap \Upsilon'_k \cap \Upsilon''_{k_2}$ we have*

$$\sum_{A<|\boldsymbol{s}|<B} |\boldsymbol{s}|^{2k} \hat{a}^2_{\boldsymbol{s}} \geqslant (1-\mu) \sum_{\boldsymbol{s} \in \mathbb{Z}^n} |\boldsymbol{s}|^{2k} \hat{a}^2_{\boldsymbol{s}}. \tag{4.6}$$

*Proof.* For any real $A > 0$ and $B > 0$ we have

$$\sum_{|\boldsymbol{s}| \leqslant A} |\boldsymbol{s}|^{2k} \hat{a}^2_{\boldsymbol{s}} \leqslant A^{2k-2k_1} \sum_{|\boldsymbol{s}| \leqslant A} |\boldsymbol{s}|^{2k_1} \hat{a}^2_{\boldsymbol{s}} \leqslant A^{2k-2k_1} \sum_{\boldsymbol{s} \in \mathbb{Z}} |\boldsymbol{s}|^{2k_1} \hat{a}^2_{\boldsymbol{s}}$$

and

$$\sum_{|\boldsymbol{s}| \geqslant B} |\boldsymbol{s}|^{2k} \hat{a}^2_{\boldsymbol{s}} \leqslant B^{2k-2k_2} \sum_{|\boldsymbol{s}| \geqslant B} |\boldsymbol{s}|^{2k_2} \hat{a}^2_{\boldsymbol{s}} \leqslant B^{2k-2k_2} \sum_{\boldsymbol{s} \in \mathbb{Z}} |\boldsymbol{s}|^{2k_2} \hat{a}^2_{\boldsymbol{s}}.$$

Under the condition $A \leqslant \bar{A}(\nu)$ for any $\nu \in \Upsilon''_{k_1} \cap \Upsilon'_k$ we get

$$\sum_{|\boldsymbol{s}| \leqslant A} |\boldsymbol{s}|^{2k} \hat{a}^2_{\boldsymbol{s}} \leqslant \tfrac{\mu}{2} c'_k \left(\tfrac{1}{\nu}\right)^{2p'(k)} \leqslant \tfrac{\mu}{2} \sum_{\boldsymbol{s} \in \mathbb{Z}} |\boldsymbol{s}|^{2k} \hat{a}^2_{\boldsymbol{s}}.$$

Under the condition $B \geqslant \bar{B}(\nu)$ for any $\nu \in \Upsilon'_k \cap \Upsilon''_{k_2}$ we get

$$\sum_{|\boldsymbol{s}| \geqslant B} |\boldsymbol{s}|^{2k} \hat{a}^2_{\boldsymbol{s}} \leqslant \tfrac{\mu}{2} c'_k \left(\tfrac{1}{\nu}\right)^{2p'(k)} \leqslant \tfrac{\mu}{2} \sum_{\boldsymbol{s} \in \mathbb{Z}} |\boldsymbol{s}|^{2k} \hat{a}^2_{\boldsymbol{s}}$$

and (4.6) follows. □

Due to the inequality

$$S_k \leqslant S_{k_1}^{\frac{k_2-k}{k_2-k_1}} S_{k_2}^{\frac{k-k_1}{k_2-k_1}}, \tag{4.7}$$

where $S_k = S_k(\nu) = \sum_{\boldsymbol{s}\in\mathbb{Z}^n} |\boldsymbol{s}|^{2k}\hat{a}_{\boldsymbol{s}}^2(\nu)$, it follows that $\bar{A} < \bar{B}$. If the closure of the set $\Upsilon''_{k_1} \cap \Upsilon'_k \cap \Upsilon''_{k_2}$ contains zero, then the powers $p''$ and $p'$ in (4.3) and (4.4) satisfy the following convexity property

$$\frac{p'(k) - p''(k_1)}{k - k_1} \leqslant \frac{p''(k_2) - p'(k)}{k_2 - k} \tag{4.8}$$

or, equivalently,

$$p'(k) \leqslant \frac{(k - k_1)p''(k_2) + (k_2 - k)p''(k_1)}{k_2 - k_1}. \tag{4.9}$$

The above two lemmas will be used to obtain bounds for the distribution of the Fourier coefficients of a function $\boldsymbol{u}$ such that $\|\boldsymbol{u}\|_k^2 \in [c'_k\nu^{-2p'(k)}, c''_k\nu^{-2p''(k)}]$ for each $k$.

We believe that for solutions of many types of PDE's and, in particular, for solutions of (1.1) we have $p'(k) = p''(k)$ and in (4.8) and (4.9) we have equalities. According to numerics of D. Jefferson (see [6]) this is the case for the complex Ginzburg–Landau equation. Moreover, for 1D Burgers-type equations this result is proven analytically (see [1]). In the case of equality in (4.8) and (4.9), lemma 5 allows to write the lower estimate for the narrowest (in terms of powers of viscosity) layer of the wave-numbers. Moreover, there is an upper bound for the sum over the same layer which coincides (in terms of powers of the viscosity) with the lower bounds. Indeed, using (4.7) we obtain

$$\sum_{\bar{A}<|\boldsymbol{s}|<\bar{B}} |\boldsymbol{s}|^{2k}\hat{a}_{\boldsymbol{s}}^2 \leqslant \sum_{\boldsymbol{s}\in\mathbb{Z}^n} |\boldsymbol{s}|^{2k}\hat{a}_{\boldsymbol{s}}^2 \leqslant (c''_{k_1})^{\frac{k_2-k}{k_2-k_1}} (c''_{k_2})^{\frac{k-k_1}{k_2-k_1}} \left(\tfrac{1}{\nu}\right)^{2p'(k)},$$

while by lemma 5 we have

$$\sum_{\bar{A}<|\boldsymbol{s}|<\bar{B}} |\boldsymbol{s}|^{2k}\hat{a}_{\boldsymbol{s}}^2 \geqslant (1-\mu)c'_k \left(\tfrac{1}{\nu}\right)^{2p'(k)}.$$

Under assumptions of Lemma 5 we cannot, in general, expect any lower bound outside the layer $\bar{A} < |\boldsymbol{s}| < \bar{B}$. Indeed, let $\gamma > 0$ and $b$ be any real numbers and suppose that $S_k(\nu) \leqslant c''_k\left(\frac{1}{\nu}\right)^{2k\gamma-2b}$ and $S_k(\nu) \geqslant c'_k\left(\frac{1}{\nu}\right)^{2k\gamma-2b}$. Then the coefficients $\hat{a}_{\boldsymbol{s}}^2(\nu)$ could be as follows:

$$\hat{a}_{\boldsymbol{s}}^2(\nu) = \begin{cases} \nu^{2b} & \text{for } \boldsymbol{s} = ([\frac{1}{\nu^\gamma}], 0, \dots, 0), \\ 0 & \text{otherwise.} \end{cases}$$

To connect the results obtained with the turbulence theory we give a *possible* rigorous definition of K–O type law and then obtain bounds for two the most important parameters of the law.

**Definition 4.** We say that positive quantities $\hat{a}_{\boldsymbol{s}}^2(\nu)$ obey a *K-O type spectral law* if there exist positive real numbers $\nu_0$, $\kappa_1 \leqslant \kappa_2$, $\varkappa$, $c$, $C$, $C_1$ positive real functions $\sigma_1(\nu)$ and $\sigma_2(\nu)$ such that $\log(\sigma_i(\nu)) = \bar{\bar{o}}(\log(\nu^{-1}))$ as $\nu \to 0$ (and $\sup \frac{\sigma_1(\nu)}{\sigma_2(\nu)} < 1$ if $\kappa_1 = \kappa_2$) such that for any $\nu \in (0, \nu_0)$ we have

1. $\sum_{r-C_1<|\boldsymbol{s}|<r+C_1} \hat{a}^2_{\boldsymbol{s}}(\nu) \leqslant C\left(\frac{1}{r}\right)^{\varkappa}$ for $r > \sigma_1(\nu)\left(\frac{1}{\nu}\right)^{\kappa_1}$
2. $\sum_{r-C_1<|\boldsymbol{s}|<r+C_1} \hat{a}^2_{\boldsymbol{s}}(\nu) \geqslant c\left(\frac{1}{r}\right)^{\varkappa}$ for $\sigma_1(\nu)\left(\frac{1}{\nu}\right)^{\kappa_1} < r < \sigma_2(\nu)\left(\frac{1}{\nu}\right)^{\kappa_2}$
3. There exist function $\varphi(k) = \bar{\bar{o}}(k)$ as $k \to \infty$ and positive functions $C(k)$, $\nu(k)$ such that for any sufficiently large $k$, any $r > 0$, and any $\nu \in (0, \nu(k))$ we have
$$\sum_{|\boldsymbol{s}|\geqslant r} \hat{a}^2_{\boldsymbol{s}}(\nu) \leqslant C(k)\left(r\nu^{\kappa_2}\right)^{-k} \nu^{-\varphi(k)}. \tag{4.10}$$

We will interpret the quantities $\hat{a}^2_{\boldsymbol{s}}(\nu)$ as the Fourier coefficients of a family of functions depending on the parameter $\nu$. Then the first and the second conditions means that the $L_2$-norm that is carried by the modes on the sphere of radius $r$, up to constants, behaves as $r^{-\varkappa}$ for wave numbers $r$ from the inertial range $r_1(\nu) < r < r_2(\nu)$. While for $r \geqslant r_2(\nu)$ we have an upper bound only. The third condition, in particular, means that for fixed $\nu$ the sum $\sum_{|\boldsymbol{s}|\geqslant r} \hat{a}^2_{\boldsymbol{s}}(\nu)$ becomes small for $r > \nu^{-\kappa_2}$ and decays faster than any finite power of $\frac{1}{r}$. Also it says that for any $\varepsilon > 0$ and $M > 0$ the quantity $\sum_{|\boldsymbol{s}|\geqslant \nu^{-\kappa_2-\varepsilon}} \hat{a}^2_{\boldsymbol{s}}(\nu)$ decays faster than $\nu^M$ as $\nu \to 0$.

Below we give a simple sufficient condition which implies (4.10). This condition covers the case of linear dependence of the power of the viscosity in the upper bounds for Sobolev norm (see (4.1)) on the its number.

**Proposition 5.** *Suppose that for any $k > 0$ and any $\nu \in (0,1)$ we have $\sum |\boldsymbol{s}|^{2k} \hat{a}^2_{\boldsymbol{s}}(\nu) \leqslant c''_k \nu^{-2\kappa_2 k+q}$. Then (4.10) holds with $C(k) = c''_{k/2}$, $\varphi(k) = q$ and $\nu(k) = 1$.*

*Proof.* Follows from Lemma 4. □

**Lemma 6.** *Suppose that the quantities $\hat{a}^2_{\boldsymbol{s}}(\nu)$ obey a K-O type spectral law in the sense of Definition 4. Then the values $\varkappa$ and $\kappa_2$ are uniquely defined.*

*Proof.* The statement about $\varkappa$ is obvious since the interval $(\nu^{-\kappa_1}\sigma_1(\nu), \nu^{-\kappa_2}\sigma_2(\nu))$ is non-empty for small $\nu$. The second statement follows from the relation:
$$\kappa_2 = \inf\{\kappa : \forall M \text{ the sum } \sum_{|\boldsymbol{s}|>\nu^{-\kappa}} \hat{a}^2_{\boldsymbol{s}}(\nu) \text{ decays faster than } \nu^M \text{ as } \nu \to 0\}. \tag{4.11}$$
□

The number $\varkappa$ is called the exponent of the K–O law and the number $\kappa_2$ is the power of the Kolmogorov dissipative scale. The quantity $\nu^{\kappa_2}$ is the Kolmogorov dissipative scale.

The value $\kappa_1$ (as well as $c$, $C$, $C_1$ and $\nu_0$) is not, in general, uniquely defined. For example if $\kappa_1 < \kappa_2$ then we can replace $\kappa_1$ with any real number in $(\kappa_1, \kappa_2]$.

**Lemma 7.** *Let us assume that we are given:*

i) *real numbers $p$, $z$ and $0 < h_1 \leqslant h_2$;*

ii) *positive real functions* $\sigma_1(\nu)$ *and* $\sigma_2(\nu)$ *such that* $\log(\sigma_i(\nu)) = \bar{\bar{o}}(\log(\nu^{-1}))$ *as* $\nu \to 0$ (*and* $\sup \frac{\sigma_1(\nu)}{\sigma_2(\nu)} < 1$ *if* $h_1 = h_2$).

*Suppose that the inequalities*

a) $$\int_{\sigma_1(\nu)\,\nu^{-h_1}}^{\sigma_2(\nu)\,\nu^{-h_2}} x^z\,dx \leqslant C''\nu^{-p}$$ b) $$\int_{\sigma_1(\nu)\,\nu^{-h_1}}^{\sigma_2(\nu)\,\nu^{-h_2}} x^z\,dx \geqslant C'\nu^{-p}$$

*hold with some positive constants* $C''$ *and* $C'$ *for a set* $\Upsilon$ *of values* $\nu$ *that contains 0 in its closure. Then*

a)
$$\begin{array}{ll} z \leqslant \frac{p}{h_2} - 1 & \text{for } p > 0, \\ z \leqslant -1 & \text{for } p = 0, \\ z < -1 & \text{for } p = 0 \text{ and } h_1 < h_2, \\ z \leqslant \frac{p}{h_1} - 1 & \text{for } p < 0, \end{array}$$

b)
$$\begin{array}{ll} z \geqslant \frac{p}{h_2} - 1 & \text{for } p > 0, \\ z \geqslant -1 & \text{for } p = 0, \\ z \geqslant \frac{p}{h_1} - 1 & \text{for } p < 0. \end{array}$$

*Proof.* For the brevity, we write $\sigma_1$ and $\sigma_2$ for $\sigma_1(\nu)$ and $\sigma_2(\nu)$. It is clear that for small enough $\nu$ (such that $1 < \sigma_1\nu^{-h_1} < \sigma_2\nu^{-h_2}$) the integral $\int_{\sigma_1\nu^{-h_1}}^{\sigma_2\nu^{-h_2}} x^z\,dx$ increases with $z$. The rest of the proof follows from the following direct calculations:

$$\int_{\sigma_1\nu^{-h_1}}^{\sigma_2\nu^{-h_2}} x^z\,dx = \begin{cases} \frac{1}{z+1}\sigma_2^{z+1}\nu^{-h_2(z+1)}\left(1 - \left(\frac{\sigma_1}{\sigma_2}\right)^{z+1}\nu^{(h_2-h_1)(z+1)}\right) & \text{if } z > -1, \\ (h_2 - h_1)\log(\nu^{-1}) + \log(\sigma_2/\sigma_1) & \text{if } z = -1, \\ \frac{-1}{z+1}\sigma_1^{z+1}\nu^{-h_1(z+1)}\left(1 - \left(\frac{\sigma_2}{\sigma_1}\right)^{z+1}\nu^{(h_1-h_2)(z+1)}\right) & \text{if } z < -1. \end{cases}$$

□

**Lemma 8.** a) *Under the assumptions of Lemma 4 suppose that* $\overline{\Upsilon''_k} \ni 0$ *and that the quantities* $\hat{a}^2_{\boldsymbol{s}}(\nu)$ *obey a K–O type spectral law. Then*

$$\begin{array}{ll} \varkappa \geqslant -\frac{2p''(k)}{\kappa_2} + 2k + 1 & \text{if } p''(k) > 0, \\ \varkappa \geqslant 1 + 2k & \text{if } p''(k) = 0, \\ \varkappa > 1 + 2k & \text{if } p''(k) = 0 \text{ and } \kappa_1 < \kappa_2, \\ \varkappa \geqslant -\frac{2p''(k)}{\kappa_1} + 2k + 1 & \text{if } p''(k) < 0. \end{array}$$

b) *Suppose that there are real sequences* $\{k_i\}$ *and* $\{\nu(k_i)\}$, $k_i \to \infty$, *such that Lemma 4 holds for any* $k = k_i$ *with* $\Upsilon''_{k_i} = (0, \nu(k_i))$. *Then*

$$\kappa_2 \leqslant \liminf \frac{p''(k_i)}{k_i}. \tag{4.12}$$

*Proof.* a) By condition 2 of Definition 4 we have

$$\sum_{r-C_1<|\boldsymbol{s}|<r+C_1} |\boldsymbol{s}|^{2k}\hat{a}^2_{\boldsymbol{s}}(\nu) \geqslant const\, r^{2k-\varkappa} \quad \text{for} \quad \sigma_1(\nu)\left(\tfrac{1}{\nu}\right)^{\kappa_1} < r < \sigma_2(\nu)\left(\tfrac{1}{\nu}\right)^{\kappa_2}.$$

Hence (with a different constant) we have

$$\sum_{\sigma_1(\nu)\,\nu^{-\kappa_1}<|\boldsymbol{s}|<\sigma_1(\nu)\,\nu^{-\kappa_1}} |\boldsymbol{s}|^{2k}\hat{a}_{\boldsymbol{s}}^2(\nu) \geqslant const \int_{\sigma_1\nu^{-\kappa_1}}^{\sigma_2\nu^{-\kappa_2}} x^{2k-\varkappa}dx.$$

Using assumption (4.1) we obtain the following inequality (again with a different constant)

$$\int_{\sigma_1\nu^{-\kappa_1}}^{\sigma_2\nu^{-\kappa_2}} x^{2k-\varkappa}dx \leqslant const\,\nu^{-2p''(k)}.$$

Now we apply Lemma 7.a with $z = 2k - \varkappa$, $p = 2p''(k)$, $h_1 = \kappa_1$, $h_2 = \kappa_2$, and $\Upsilon = \Upsilon_k'' \cap (0, \nu_0)$ to complete the proof of the first part of the lemma.

b) We prove that for any $\gamma \geqslant \liminf \frac{p''(k_i)}{k_i}$ condition 3 of Definition 4 holds if the value $\kappa_2$ in (4.10) is replaced with $\gamma$.

To do this we take a subsequence $\{k_{j_i}\}$ such that $\liminf \frac{p''(k_i)}{k_i} = \lim \frac{p''(k_{j_i})}{k_{j_i}}$. Then

$$\max\{0, \tfrac{p''(k_{j_i})}{k_{j_i}} - \gamma\} = \bar{\bar{o}}(1) \text{ as } i \to \infty. \tag{4.13}$$

Now applying (4.2) for each $k = k_{j_i}$ we obtain

$$\sum_{|\boldsymbol{s}|\geqslant r} \hat{a}_{\boldsymbol{s}}^2(\nu) \leqslant C(2k_{j_i})\,(r\nu^{\gamma})^{-2k_{j_i}}\,\nu^{-\varphi(2k_{j_i})} \quad \text{for any } i \text{ and any } \nu \in (0, \nu(k_{j_i})). \tag{4.14}$$

where $C(2k_{j_i}) = c''_{k_{j_i}}$ and $\varphi(2k_{j_i}) = \max\{0, 2p''(k_{j_i}) - 2k_{j_i}\gamma\}$ ($= \bar{\bar{o}}(2k_{j_i})$ due to (4.13)). Using the interpolation we see that (4.14) holds when $2k_{j_i}$ is replaced with any $k \geqslant 0$ (and the functions $C(\cdot)$ and $\varphi(\cdot)$ are determined according to this interpolation). It follows that $\kappa_2 \leqslant \gamma$. Hence $\kappa_2 \leqslant \liminf \frac{p''(k_i)}{k_i}$. □

**Lemma 9.** a) *Under the assumptions of Lemma* 5 *suppose that* $\overline{\Upsilon_{k_1}'' \cap \Upsilon_k' \cap \Upsilon_{k_2}''} \ni 0$, *and that the quantities* $\hat{a}_{\boldsymbol{s}}^2(\nu)$ *obey a K–O type spectral law. Then*

$$\kappa_2 \geqslant \frac{p'(k) - p''(k_1)}{k - k_1}. \tag{4.15}$$

b) *If, in addition, we have* $\kappa_1 < \frac{p'(k)-p''(k_1)}{k-k_1}$ *then*

$$\varkappa \leqslant -\frac{2p'(k)}{\min\left\{\kappa_2, \frac{p''(k_2)-p'(k)}{k_2-k}\right\}} + 2k + 1 \quad \textit{if } p'(k) > 0,$$

$$\varkappa \leqslant 1 + 2k \quad \textit{if } p'(k) = 0,$$

$$\varkappa \leqslant -2p'(k)\frac{k-k_1}{p'(k)-p''(k_1)} + 2k + 1 \quad \textit{if } p'(k) < 0.$$

*Proof.* a) We denote $p_- = \frac{p'(k)-p''(k_1)}{k-k_1}$ and $p_+ = \frac{p''(k_2)-p'(k)}{k_2-k}$. To prove (4.15) we show that for any $\varepsilon > 0$ the sum $\sum_{|\boldsymbol{s}|>\nu^{-(p_- -\varepsilon)}} \hat{a}_{\boldsymbol{s}}^2(\nu)$ decays as $\nu \to 0$ not faster than some finite power of $\nu$.

Fix any $\varepsilon > 0$. Then for small enough $\nu$ such that $\nu^{-(p_- - \varepsilon)} < \bar{A}(\nu)$ (the quantities $\bar{A}(\nu)$ and $\bar{B}(\nu)$ are defined by (4.5) with $\mu = 1/2$) due to Lemma 5 we have

$$\sum_{|\boldsymbol{s}| > \nu^{-(p_- - \varepsilon)}} \hat{a}_{\boldsymbol{s}}^2(\nu) \geqslant \sum_{\bar{A}(\nu) < |\boldsymbol{s}| < \bar{B}(\nu)} \hat{a}_{\boldsymbol{s}}^2(\nu)$$
$$\geqslant (\bar{B}(\nu))^{-2k} \sum_{\bar{A}(\nu) < |\boldsymbol{s}| < \bar{B}(\nu)} |\boldsymbol{s}|^{2k} \hat{a}_{\boldsymbol{s}}^2(\nu) \geqslant const\, \nu^{2kp_+ - 2p'(k)}.$$

b) Let $\bar{\bar{B}}_\varepsilon(\nu) = \min\{\bar{B}(\nu), \nu^{-(\kappa_2 + \varepsilon)}\}$. First we claim that for small enough $\nu$ we have

$$\sum_{\bar{A}(\nu) < |\boldsymbol{s}| < \bar{\bar{B}}_\varepsilon(\nu)} |\boldsymbol{s}|^{2k} \hat{a}_{\boldsymbol{s}}^2 \geqslant const\, \nu^{-2p'(k)}.$$

Indeed, if $\bar{B}(\nu) \leqslant \nu^{-(\kappa_2 + \varepsilon)}$, then the inequality directly follows from Lemma 5. If $\bar{B}(\nu) > \nu^{-(\kappa_2 + \varepsilon)}$, then it follows from Lemma 5 and the fact that the sum $\sum_{|\boldsymbol{s}| > \nu^{-(\kappa_2 + \varepsilon)}} \hat{a}_{\boldsymbol{s}}^2(\nu)$ decays faster then any power of $\nu$.

The assumption $\kappa_1 < p_-$ implies that for small enough $\nu$ we have $\sigma_1(\nu)\nu^{-\kappa_1} < \bar{A}(\nu)$ and so, by the condition 1 of Definition 4, we have

$$\sum_{\bar{A}(\nu) < |\boldsymbol{s}| < \bar{\bar{B}}_\varepsilon(\nu)} |\boldsymbol{s}|^{2k} \hat{a}_{\boldsymbol{s}}^2 \leqslant const \int_{\bar{A}(\nu)}^{\bar{\bar{B}}_\varepsilon(\nu)} x^{2k - \varkappa} dx.$$

This inequality and the claim above imply

$$\int_{\bar{A}(\nu)}^{\bar{\bar{B}}_\varepsilon(\nu)} x^{2k - \varkappa} dx \geqslant const\, \nu^{-2p'(k)}.$$

Now we apply Lemma 7.b and take the limit $\varepsilon \to 0$ to complete the proof of the second part of the lemma. □

### 4.2. Back to the Burgers equation

Here we apply the results of the previous subsection to estimate distribution of the averaged Fourier coefficients for solutions of (1.1).

For the sequel we note that if

$$q_n = \left[\tfrac{n}{2}\right] + 1, \tag{4.16}$$

then for any vector-valued function $\boldsymbol{u}$ on the torus $\mathbb{T}^n$, any $k \geqslant q_n + 1$ and any index $j$ we have

$$\|\boldsymbol{u}\|_k^2 \geqslant K_n^2 \Big| \frac{\partial^{k - q_n} \boldsymbol{u}}{\partial x_j^{k - q_n}} \Big|_{L_\infty}^2. \tag{4.17}$$

where

$$K_n = \ell^{n/2} \left(\tfrac{2\pi}{\ell}\right)^{2q_n} \Big( \sum_{\boldsymbol{s} \in \mathbb{Z}^n \setminus \{0\}} \tfrac{1}{|\boldsymbol{s}|^{2q_n}} \Big)^{-\frac{1}{2}}. \tag{4.18}$$

**Lemma 10.** *Let $\boldsymbol{u} = \boldsymbol{u}^\nu(t,\boldsymbol{x})$ be a solution of (1.1) with an initial state $\boldsymbol{u}_0$.*

i) *Let $\nu_0$ and $T$ be any positive numbers. Then for each $k \geqslant 0$ there exists $\widetilde{C}_k = \widetilde{C}_k(T,\nu_0)$ such that for any positive $\nu < \nu_0$ we have*

$$\frac{1}{T}\int_0^T \|\boldsymbol{u}\|_k^2 dt \leqslant \widetilde{C}_k \left(\tfrac{1}{\nu}\right)^{2k}. \tag{4.19}$$

ii) *Suppose that $\boldsymbol{u}_0$ is non-degenerate. Let $T = T(\boldsymbol{f},\boldsymbol{u}_0)$ be given by (3.16). Suppose that the forcing term $\boldsymbol{h}$ in (1.1) satisfies the condition $H(T) < \frac{c}{2}$, where $c$ is given by (3.33). Then for any integer $k \geqslant 2 + q_n$ and for any $\nu > 0$, we have*

$$\frac{1}{T}\int_0^T \|\boldsymbol{u}\|_k^2 dt \geqslant K_n^2 r_{k-q_n}^2 \left(\tfrac{1}{\nu}\right)^{k-q_n}. \tag{4.20}$$

*Here $r_k$, $q_n$ and $K_n$ is given by (3.11), (4.16) and (4.18), respectively.*

*Proof.* i) Using (1.6) we arrive at (4.19) with
$\widetilde{C}_k = R_k^2(T)\max\{1, \frac{\|\boldsymbol{u}_0\|_k^2}{R_k(0)^2}\nu_0^{2k}, \frac{\sup_{[0,t]}\|\boldsymbol{h}\|_{k-1}^2}{R_k(0)^2}\nu_0^{2k}\}$.

ii) Using inequality (4.17), the Cauchy–Schwartz inequality and Theorem 2 we obtain:

$$\frac{1}{T}\int_0^T \|\boldsymbol{u}\|_k^2 dt \geqslant K_n^2 \frac{1}{T}\int_0^T \left|\frac{\partial^{k-q_n}\boldsymbol{u}}{\partial x_j^{k-q_n}}\right|_{L_\infty}^2 dt \geqslant K_n^2 \left(\frac{1}{T}\int_0^T \left|\frac{\partial^{k-q_n}\boldsymbol{u}}{\partial x_j^{k-q_n}}\right|_{L_\infty} dt\right)^2 \geqslant K_n^2 \frac{r_{k-q_n}^2}{\nu^{k-q_n}}.$$

□

Consider the orthonormal basis on $L_2(\mathbb{T}^n)$, formed by the exponents $e_{\boldsymbol{s}}(\cdot)$, $\boldsymbol{s} \in \mathbb{Z}^n$, where

$$e_{\boldsymbol{s}}(\boldsymbol{x}) = \tfrac{1}{\ell^{n/2}} \exp\left(\tfrac{2\pi i \boldsymbol{s}\boldsymbol{x}}{\ell}\right).$$

The Fourier expansion of $\boldsymbol{u}(\boldsymbol{x})$ has the form

$$\boldsymbol{u}(t,\boldsymbol{x}) = \sum_{\boldsymbol{s}\in\mathbb{Z}^n} \hat{\boldsymbol{u}}_{\boldsymbol{s}}(t) e_{\boldsymbol{s}}(\boldsymbol{x}), \quad \text{where} \quad \hat{\boldsymbol{u}}_{\boldsymbol{s}}(t) = \int_{\mathbb{T}^n} \boldsymbol{u}(t,\boldsymbol{x})\overline{e_{\boldsymbol{s}}(\boldsymbol{x})} d\boldsymbol{x}.$$

For the Sobolev quasinorm $\|\cdot\|_k$ (1.5) we have $\|\boldsymbol{u}\|_k^2 = \left(\frac{2\pi}{\ell}\right)^{2k} \sum_{\boldsymbol{s}\in\mathbb{Z}^n} |\boldsymbol{s}|^{2k} |\hat{\boldsymbol{u}}_{\boldsymbol{s}}|^2$.

Suppose that the initial state $\boldsymbol{u}_0$ is a non-degenerate vector field. Consider the corresponding solution of equation (1.1) and define the quantities $\hat{a}_{\boldsymbol{s}}^2(\nu)$ by the formula

$$\hat{a}_{\boldsymbol{s}}^2 = \frac{1}{T}\int_0^T |\hat{\boldsymbol{u}}_{\boldsymbol{s}}(t)|^2 dt, \tag{4.21}$$

where $T = T(\boldsymbol{f},\boldsymbol{u}_0)$ is defined by (3.16). Then we have

$$\sum_{\boldsymbol{s}\in\mathbb{Z}^n} |\boldsymbol{s}|^{2k} \hat{a}_{\boldsymbol{s}}^2(\nu) = \left(\frac{\ell}{2\pi}\right)^{2k} \frac{1}{T}\int_0^T \|\boldsymbol{u}^\nu\|_k^2\, dt.$$

Lemma 10 implies that for any $k$ and any $\nu \in (0,\nu_0)$ we have the following inequality

$$\sum_{\boldsymbol{s}\in\mathbb{Z}^n} |\boldsymbol{s}|^{2k} \hat{a}_{\boldsymbol{s}}^2(\nu) \leqslant \left(\tfrac{\ell}{2\pi}\right)^{2k} \widetilde{C}_k \left(\tfrac{1}{\nu}\right)^{2k}. \tag{4.22}$$

Under the assumptions of the second part of Lemma 10 we see that for any $\nu > 0$ and any integer $k \geqslant q_n + 2$ there is the inequality:

$$\sum_{\boldsymbol{s}\in\mathbb{Z}^n} |\boldsymbol{s}|^{2k} \hat{a}_{\boldsymbol{s}}^2(\nu) \geqslant K_n^2 r_{k-q_n}^2 \left(\tfrac{\ell}{2\pi}\right)^{2k} \left(\tfrac{1}{\nu}\right)^{k-q_n}. \tag{4.23}$$

We recall that $q_n = \left[\frac{n}{2}\right] + 1$ (4.16).

**Theorem 6.** *Let $\boldsymbol{u} = \boldsymbol{u}^\nu(t, \boldsymbol{x})$ be the solution of (1.1) with a non-degenerate initial state $\boldsymbol{u}_0$. Take $T = T(\boldsymbol{f}, \boldsymbol{u}_0)$ by (3.16) and consider the averaged Fourier coefficients $\hat{a}_{\boldsymbol{s}}^2(\nu)$ given by (4.21). Suppose that the forcing term $\boldsymbol{h}$ in (1.1) satisfies the condition $H(T) < \frac{c}{2}$, where c is given by the right hand side of (3.33). Then for any $k \geqslant 2 + q_n$, where $q_n$ is given by (4.16), there exists $c_k$ such that for any $\varepsilon > 0$ and any small enough $\nu$ we have*

$$\sum_{(\frac{1}{\nu})^{\frac{1}{2}-\frac{q_n}{2k}-\varepsilon} < |\boldsymbol{s}| < (\frac{1}{\nu})^{1+\varepsilon}} \hat{a}_{\boldsymbol{s}}^2(\nu) \geqslant const_k \nu^{k+2k\varepsilon+q_n}. \tag{4.24}$$

*For any $k \geqslant 0$ and $\nu_0 > 0$ there exists $C_k = C_k(\nu_0)$ such that for any positive real numbers $z$ and $\lambda_1 < \lambda_2$, and any $\nu < \nu_0$ we have*

$$\sum_{\lambda_1(\frac{1}{\nu})^z \leqslant |\boldsymbol{s}| \leqslant \lambda_2(\frac{1}{\nu})^z} \hat{a}_{\boldsymbol{s}}^2(\nu) \leqslant \lambda_1^{-2k} C_k\, \nu^{2k(z-1)}. \tag{4.25}$$

*Besides, for any $y > 0$ we have*

$$\sum_{|\boldsymbol{s}|\geqslant y} \hat{a}_{\boldsymbol{s}}^2(\nu) \leqslant C_k (y\nu)^{-2k}. \tag{4.26}$$

*Proof.* The quantities $\hat{a}_s^2(\nu)$ satisfy inequalities (4.23) and (4.22). Hence we can apply Lemmas 4 and 5 with $p''(k) = k$, $p'(k) = k/2 - q_n/2$ for $k \geqslant q_n + 2$, $c_k'' = \left(\frac{\ell}{2\pi}\right)^{2k} \widetilde{C}_k$ $c_k' = \left(\frac{\ell}{2\pi}\right)^{2k} K_n^2 r_{k-q_n}^2$. Choose $k_1 = 0$ and $k_2 > k$ such that

$$\frac{k_2 - (k/2 - q_n/2)}{k_2 - k} < 1 + \varepsilon,$$

and $\nu(k, \varepsilon, c_i', c_i'')$ such that for any positive $\nu < \nu(k, \varepsilon, c_i', c_i'')$ we have

$$\left(\tfrac{1}{\nu}\right)^{\frac{1}{2}-\frac{q_n}{2k}-\varepsilon} < \bar{A}(\nu) \quad \text{and } \bar{B}(\nu) > \left(\tfrac{1}{\nu}\right)^{1+\varepsilon},$$

where $\bar{A}(\nu)$ and $\bar{B}(\nu)$ are defined by (4.5) with $\mu = 1/2$. Now we apply Lemma 5 to obtain

$$\sum_{(\frac{1}{\nu})^{\frac{1}{2}-\frac{q_n}{2k}-\varepsilon} < |\boldsymbol{s}| < (\frac{1}{\nu})^{1+\varepsilon}} |\boldsymbol{s}|^{2k} \hat{a}_{\boldsymbol{s}}^2 \geqslant \tfrac{1}{2} c_k' \left(\tfrac{1}{\nu}\right)^{k-q_n}.$$

Using the inequality

$$\sum_{(\frac{1}{\nu})^{\frac{1}{2}-\frac{q_n}{2k}-\varepsilon} < |\boldsymbol{s}| < (\frac{1}{\nu})^{1+\varepsilon}} \hat{a}_{\boldsymbol{s}}^2 \geqslant \nu^{2k(1+\varepsilon)} \sum_{(\frac{1}{\nu})^{\frac{1}{2}-\frac{q_n}{2k}-\varepsilon} < |\boldsymbol{s}| < (\frac{1}{\nu})^{1+\varepsilon}} |\boldsymbol{s}|^{2k} \hat{a}_{\boldsymbol{s}}^2$$

we arrive at (4.24) with $const_k = c_k'/2$.

Using inequality (4.22) and Lemma 4 we arrive at (4.25) and (4.26) with $C_k(\nu_0) = c''_k$. □

Under the assumptions of Theorem 6, we have the following conditional result:

**Theorem 7.** *If the quantities $\hat{a}_s^2(\nu)$, defined by (4.21), obey a K-O type spectral law, then the power of the Kolmogorov dissipation scale $\kappa_2 \in [1/2, 1]$. For the power of the spectral law $\varkappa$ we have $\varkappa \geqslant 1$, and $\varkappa > 1$ if $\kappa_1 < \kappa_2$.*

*Suppose, in addition, that the energy range is small enough, that is $\kappa_1 < \frac{1}{2} - \frac{q_n}{2k}$ for some integer $k \geqslant q_n + 2$; then we have $\varkappa \leqslant k + q_n + 1$. In particular, if $\kappa_1 < \frac{1}{q_n+2}$ then we have $\varkappa \leqslant 2q_n + 3$.*

*Proof.* For any $k \geqslant 0$ the quantities $\hat{a}_s^2(\nu)$ satisfy inequalities (4.1) with $p''(k) = k$, while for $k \geqslant q_n + 2$ the quantities $\hat{a}_s^2(\nu)$ satisfy inequalities (4.4) with $p'(k) = \frac{1}{2}(k - q_n)$. By (4.12) we have $\kappa_2 \leqslant 1$. Taking any $k_1$ in (4.15) and then passing $k \to \infty$ we obtain $\kappa_2 \geqslant 1/2$. Using Lemma 8.a with $k = 0$ we obtain that $\varkappa \geqslant 1$ and $\varkappa > 1$ if $\kappa_1 < \kappa_2$. If $\kappa_1 < \frac{p'(k)-p''(0)}{k-0}$ and $k \geqslant q_n + 2$, then by Lemma 9.b we have

$$\varkappa \leqslant 2k + 1 - \frac{k-q_n}{\kappa_2}$$

Using the inequality $\kappa_2 \leqslant 1$ we obtain $\varkappa \leqslant k + 1 + q_n$. If this inequality holds for $k = q_n + 2$, i.e., if $\kappa_1 < \frac{p'(q_n+2)}{q_n+2} = \frac{1}{q_n+2}$, then we have $\varkappa \leqslant 2q_n + 3$. □

## 5. Low bounds for spatial derivatives of solutions of the Navier–Stokes system

In [2] low bounds of the $L_\infty$ time-space norm of solutions of the $nD$ free Navier–Stokes system are obtained. In this section we amplify those estimates to the case of $L_1$ in time of the $L_\infty$ in space norm.

The free Navier–Stokes system can be regarded as a special case of the equation (1.1) when $m = n$, $\boldsymbol{f}(\boldsymbol{u}) = \boldsymbol{u}$ and $\boldsymbol{h}$ represent the gradient of the pressure term. This observation allows us to extend the results of previous sections to solutions of the Navier–Stokes system.

In this section we consider the dynamics of a vector field $\boldsymbol{u} = \boldsymbol{u}(t, \boldsymbol{x})$ on the torus $\mathbb{T}^n = {}^{\mathbb{R}^n}\!/_{\ell\mathbb{Z}^n}$ described by the Navier–Stokes system:

$$\partial_t \boldsymbol{u} + \nabla_{\boldsymbol{u}} \boldsymbol{u} = \nu \Delta \boldsymbol{u} + \nabla p(t, \boldsymbol{x}), \tag{5.1}$$

$$\operatorname{div} \boldsymbol{u} = 0, \tag{5.2}$$

with a positive viscosity $\nu$. We assume that the initial state $\boldsymbol{u}(0, \boldsymbol{x}) = \boldsymbol{u}_0$ is $C^2$-smooth. Let $T_0 = T_0(\boldsymbol{u}_0)$ be supremum of existence time-interval, i.e., the smooth solution of the system (5.1),(5.2) exists for $t \in [0, T_0)$. There are lower estimates for $T_0$; for example, in the paper [11] it is shown that $T_0 \geqslant \frac{C(n)\,\ell^2}{\nu R^2 (1+\max\{0, \log R\})^2}$, where $R = \frac{\ell}{\nu} |\boldsymbol{u}_0|_{L_\infty}$ is the Reynolds number. It is clear that if $T_0 < \infty$, then $|\boldsymbol{u}(t,\cdot)|_{L_\infty} \to \infty$ as $t \to T_0$.

**Remark 1.** Since the mean value of $\boldsymbol{u}$ is constant, then the spatial derivatives of $\boldsymbol{u}$ blow up with $|\boldsymbol{u}(t,\cdot)|_{L_\infty}$. Due to this reason we assume that $\max\limits_{j=1,\dots,n} \left|\frac{\partial^k \boldsymbol{u}(t,\cdot)}{\partial x_j^k}\right|_{L_\infty} = \infty$ for $t \geqslant T_0$ and any $k \geqslant 1$.

**Theorem 8.** *Suppose that the initial state $\boldsymbol{u}_0$ is a non-degenerate vector field. Then there exist a $\nu$-independent positive real numbers $T$ and $\varkappa_2$, $\varkappa_3$, $\varkappa_4, \dots$ such that for any $\nu > 0$ and for any $k \geqslant 2$ we have:*

$$\max\Big\{ \max_{j=1\dots n} \frac{1}{T} \int_0^T \sup_{\boldsymbol{x}\in\mathbb{T}^n} \left|\frac{\partial^k \boldsymbol{u}}{\partial {x_j}^k}(t,\boldsymbol{x})\right| dt,\ \max_{j=1\dots n} \frac{1}{T} \int_0^T \frac{1}{\nu^{k/2}} \sup_{\boldsymbol{x}\in\mathbb{T}^n} \left|\frac{\partial^{k-1} p}{\partial x_j^{k-1}}(t,\boldsymbol{x})\right| dt \Big\} \geqslant \frac{\varkappa_k}{\nu^{k/2}}. \tag{5.3}$$

As the next result shows, the assumption of non-degeneracy cannot be removed. To show this we can restrict ourself to the 2D case since any solution of the 2D Navier–Stokes system generates a solution of a higher-dimensional system by adding zeroth components of the field $\boldsymbol{u}$ and fictitious variables. The non-degeneracy is preserved by this operation. We remark that in the 2D case the non-degeneracy is a necessary and sufficient condition for theorem 8.

**Lemma 11.** *Let $\boldsymbol{u}$ be a solution of 2D Navier–Stokes system with a degenerate initial state. Then for any $t > 0$ and $k \geqslant 0$ we have* $\max|\boldsymbol{u}(t,\boldsymbol{x})|_{C^k(\mathbb{T}^2)} \leqslant \max|\boldsymbol{u}(0,\boldsymbol{x})|_{C^k(\mathbb{T}^2)}$ *and* $\nabla p \equiv 0$.

*Proof.* In the 2D case any degenerate initial state has the form:

$$\boldsymbol{u}_0(\boldsymbol{x}) = \binom{b_2}{-b_1}\varphi_0(b_1x_1 + b_2x_2) + \binom{c_1}{c_2} \tag{5.4}$$

for a suitable function $\varphi_0(\cdot)$ and real numbers $b_1$, $b_2$, $c_1$, and $c_2$ (see Theorem 3). In this case, the solution of the Cauchy problem for (5.1), (5.2) remains of the form (5.4):

$$\boldsymbol{u}(t,\boldsymbol{x}) = \binom{b_2}{-b_1}\varphi(t, b_1x_1 + b_2x_2) + \binom{c_1}{c_2},$$

where the function $\varphi$ satisfies the following linear parabolic equation with constant coefficients:

$$\varphi_t + (b_1c_1 - b_2c_2)\varphi' = (b_1^2 + b_2^2)\nu\varphi''.$$

Therefore derivatives of $\boldsymbol{u}$ are decreasing and $\nabla p \equiv 0$. □

We note that in the 2D case the class of periodic solutions with degenerate initial state coincides with the class of periodic solutions with $\nabla p \equiv 0$. (For the Euler equation this is true for any dimension, see Theorem 4) Indeed, if the initial state is degenerate, then $\nabla p \equiv 0$ due to Lemma 11. To show the converse we apply the identity

$$\Delta p(t,x) = -2I_2(t,x), \tag{5.5}$$

where $I_k(t,x)$ are the coefficients of the characteristic polynomial of the matrix $\frac{\partial \boldsymbol{u}}{\partial \boldsymbol{x}}(t,\boldsymbol{x})$:

$$\det\big(\tfrac{\partial \boldsymbol{u}}{\partial \boldsymbol{x}} - \lambda \mathbf{1}\big) = (-\lambda)^n + (-\lambda)^{n-1} I_1(t,x) + \cdots + I_n(t,x).$$

The degeneracy of the initial state is equivalent to the condition $I_1(0,\cdot) \equiv I_2(0,\cdot) \equiv \ldots \equiv I_n(0,\cdot) \equiv 0$. The coefficient $I_1 \equiv 0$ due to (5.2).

The identity (5.5) is also valid in higher dimensions. For the proof one can take the divergence of (5.1), (5.2) and note that $\operatorname{div} \nabla_{\boldsymbol{u}} \boldsymbol{u} = -2I_2(t,x)$.

*Proof of Theorem* 8. As we will show, the assertion of the theorem holds for any $\boldsymbol{u}$ and $p$ that satisfy (5.1) (we note that equation (5.2) is not used in our calculations). First, from theorem 2 we find $T = T(\boldsymbol{u}_0)$. If $T_0 < T$, then the left hand side of (5.3) is equal to infinity due to Remark 1. Suppose $T < T_0$. Then from Theorem 2 again, we find $c = c(\boldsymbol{u}_0)$ and $r_k = r_k(\boldsymbol{u}_0)$ $(k \geqslant 2)$ such that if

$$\frac{1}{T}\int_0^T \sup_{\boldsymbol{x}\in\mathbb{T}^n} \big|\boldsymbol{\nabla} p(t,\boldsymbol{x})\big| dt < \tfrac{c}{2},$$

then for any $k \geqslant 2$ we have

$$\max_{j=1\ldots n} \frac{1}{T}\int_0^T \sup_{\boldsymbol{x}\in\mathbb{T}^n} \Big|\tfrac{\partial^k \boldsymbol{u}}{\partial {x_j}^k}(t,\boldsymbol{x})\Big|\, dt \geqslant \tfrac{r_k}{\nu^{k/2}}\,.$$

If

$$\frac{1}{T}\int_0^T \sup_{\boldsymbol{x}\in\mathbb{T}^n} \big|\boldsymbol{\nabla} p(t,\boldsymbol{x})\big| dt \geqslant \tfrac{c}{2},$$

then for any $k \geqslant 2$ we have:

$$\max_{j=1\ldots n} \frac{1}{T}\int_0^T \sup_{\boldsymbol{x}\in\mathbb{T}^n} \big|\tfrac{\partial^{k-1} p}{\partial x_j^{k-1}}(t,\boldsymbol{x})\big| dt \geqslant \tfrac{c}{2\ell^{k-1}\sqrt{n}}.$$

Inequality (5.3) is proved with the constant $\varkappa_k = \min\{r_k, \frac{c}{2\ell^{k-1}\sqrt{n}}\}$. □

## References


[1] Biryuk A. *Spectral Properties of Solutions of Burgers Equation with Small Dissipation.* Funct. Anal. Appl. **35**, no. 1 (2001), 1–15.

[2] Biryuk A. *On Spatial Derivatives of Solutions of the Navier-Stokes Equation with Small Viscosity.* Uspekhi Mat. Nauk **57** no 1. (2002), 147–148.

[3] Dubrovin B. A.; Novikov S. P.; Fomenko A. T. *Modern geometry—methods and applications.* Part I. *The geometry of surfaces, transformation groups, and fields.* Part II. *The geometry and topology of manifolds.* Graduate Texts in Mathematics 93, 104. Springer-Verlag 1984, 1985.

[4] Friedman A. *Partial differential equations of parabolic type.* Prentice-Hall, Inc., Englewood Cliffs, N.J., 1964.

[5] Frish U. *Turbulence. The legacy of A.N. Kolmogorov.* Cambridge Univ. Press, 1995.

[6] Jefferson D. *A numerical and analytical approach to turbulence in a special class of complex Ginzburg–Landau equations.* Heriot-Watt University Thesis, 2002.

[7] Halmos P. R. *Finite-dimensional vector spaces.* Springer-Verlag, New York – Heidelberg, 1974.

[8] Hartman Ph. *Ordinary Differential Equations.* John Wiley & Sons Inc., New York, 1964.

[9] Hörmander L. *Lectures on nonlinear hyperbolic differential equations.* Springer-Verlag, Berlin, 1997.

[10] Kolmogorov A. N. *On inequalities for supremums of successive derivatives of a function on an infinite interval.* Paper 40 in " Selected works of A.N. Kolmogorov, vol.1 " Moscow, Nauka 1985. Engl. translation: Kluwer, 1991.

[11] Kukavica I., Grujić Z. *Space Analyticity for the Navier-Stokes and Related Equations with Initial Data in $L^p$.* Journal of Functional Analysis **152**, no. 2 (1998), 447–466.

[12] Kuksin S. *Spectral Properties of Solutions for Nonlinear PDE's in the Turbulent Regime.* GAFA, **9**, no. 1 (1999), 141–184.

[13] Pogorelov A. V. *Extensions of the theorem of Gauss on spherical representation to the case of surfaces of bounded extrinsic curvature.* Dokl. Akad. Nauk. SSSR (N.S.) **111**, no.5 (1956), 945–947.

[14] Spivak M. *A comprehensive introduction to differential geometry. Vol. III*, Publish or Perish, Boston, Mass., 1975.

Andrei Biryuk
Independent University of Moscow
11, Bol'shoj Vlas'yevskij pyeryeulok
Moscow 121002
Russia
e-mail: `biryuk@mccme.ru`

Advances in Mathematical Fluid Mechanics, 31–51


# On the Global Well-posedness and Stability of the Navier–Stokes and the Related Equations

Dongho Chae and Jihoon Lee

**Abstract.** We study the problem of global well-posedness and stability in the scale invariant Besov spaces for the modified 3D Navier–Stokes equations with the dissipation term, $-\Delta u$ replaced by $(-\Delta)^\alpha u$, $0 \le \alpha < \frac{5}{4}$. We prove the unique existence of a global-in-time solution in $B^{\frac{5}{2}-2\alpha}_{2,1}$ for initial data having small $\dot{B}^{\frac{5}{2}-2\alpha}_{2,1}$ norm for all $\alpha \in [0, \frac{5}{4})$. We also obtain the global stability of the solutions $\dot{B}^{\frac{5}{2}-2\alpha}_{2,1}$ for $\alpha \in [\frac{1}{2}, \frac{5}{4})$. In the case $\frac{1}{2} < \alpha < \frac{5}{4}$, we prove the unique existence of a global-in-time solution in $\dot{B}^{\frac{3}{p}+1-2\alpha}_{p,\infty}$ for small initial data, extending the previous results for the case $\alpha = 1$.

**Mathematics Subject Classification (2000).** 35Q30, 76D03.

**Keywords.** Navier–Stokes equations, global well-posedness, stability.

## 1. Introduction

In this paper, we are concerned with the sub-dissipative or hyper-dissipative Navier–Stokes equations.

$$(\mathrm{SNS})_\alpha \begin{cases} \partial_t u + (u\cdot\nabla)u + (-\Delta)^\alpha u + \nabla p = 0, \mathbb{R}^3 \times \mathbb{R}_+, 0 \le \alpha < \frac{5}{4}, \\ \mathrm{div}\ u = 0, \\ u(0,x) = u_0(x), \end{cases}$$

where $u$ represents the velocity vector field, $p$ is the scalar pressure. For simplicity, we assume that the external force vanishes, but it is easy to extend our results to the case of nonzero external force. J.L. Lions [22] proved the existence of a unique regular solution provided $\alpha \ge \frac{5}{4}$. If $\alpha = 1$, then the above system reduces to the usual Navier–Stokes equations. For the Navier–Stokes equations, Kato [20] proved the existence of global solution in $C([0,\infty); L^3(\mathbb{R}^3))$ if $\|u_0\|_{L^3}$ is sufficiently small. After Kato's work [20], there were many important improvements using the scaling invariant function spaces. Especially, pioneered by Chemin [12], Cannone–Meyer–Planchon [7], initial value problems of the Navier–Stokes equations in some Besov spaces were extensively studied (see also [3], [4], [6] and [9]). For the Euler equations and compressible or incompressible Navier–Stokes equations, there are many recent improvements using the notion of the Besov spaces and Triebel–Lizorkin spaces (see [10], [11], [13], [14], [16], [17], [21], [26] and references therein). Recently, Cannone–Karch [5] proved some existence and uniqueness theorems of global-in-time solutions with external force and small initial conditions in some Besov type spaces in the case that $-\Delta+(-\Delta)^\alpha$ replaces $(-\Delta)^\alpha$. Considering scaling analysis, we find that if $u(x,t)$ is a solution of $(\mathrm{SNS})_\alpha$, then $u_\lambda(x,t)=\lambda^{2\alpha-1}u(\lambda x,\lambda^{2\alpha}t)$ is also a solution of $(\mathrm{SNS})_\alpha$. Thus $\dot B^{\frac{3}{p}+1-2\alpha}_{p,q}$, $1\le p,q\le\infty$, are scaling invariant function spaces. Our first main result of this paper is the global existence and uniqueness result for the initial value problem $(\mathrm{SNS})_\alpha$ with the initial data small in $\dot B^{\frac52-2\alpha}_{2,1}$ norm. Precise statement is as follows.

**Theorem 1.** *Let $\alpha\in[0,\frac54)$ be given. There exists a constant $\epsilon>0$ such that for any $u_0\in B^{\frac52-2\alpha}_{2,1}$ and $\|u_0\|_{\dot B^{\frac52-2\alpha}_{2,1}}<\epsilon$, the IVP $(\mathrm{SNS})_\alpha$ has a global unique solution $u$, which belongs to $L^\infty(0,\infty;B^{\frac52-2\alpha}_{2,1})\cap L^1(0,\infty;\dot B^{\frac52}_{2,1})\cap C([0,\infty);B^{\beta}_{2,1})$ with $\beta=\begin{cases}\frac52-2\alpha, & \text{if } \frac12\le\alpha\le\frac54\\ \frac52-2\alpha-\delta_1, & \text{if } 0\le\alpha<\frac12,\end{cases}$ for $\delta_1>0$. Moreover, for any $\sigma>0$, $u$ also belongs to $L^\infty(\sigma,\infty;B^{\frac52}_{2,1})\cap L^1(\sigma,\infty;\dot B^{\frac52+2\alpha}_{2,1})\cap C((0,\infty);B^{\gamma}_{2,1})$, where $\gamma=\begin{cases}\frac52-\delta_2, & \text{if } 0\le\alpha<\frac12,\\ \frac52, & \text{if } \frac12\le\alpha<\frac54,\end{cases}$ for any $\delta_2>0$. Furthermore, the solution $u$ satisfies the following estimates*

$$\sup_{0\le t<\infty}\|u(t)\|_{B^{\frac52-2\alpha}_{2,1}}+C\int_0^\infty\|u(t)\|_{\dot B^{\frac52}_{2,1}}dt\le\|u_0\|_{B^{\frac52-2\alpha}_{2,1}}\exp\left(C\int_0^\infty\|u(t)\|_{\dot B^{\frac52}_{2,1}}dt\right),$$

*and for any $\sigma>0$,*

$$\sup_{\sigma\le t<\infty}\|u(t)\|_{B^{\frac52}_{2,1}}+C\int_\sigma^\infty\|u(t)\|_{\dot B^{\frac52+2\alpha}_{2,1}}dt\le\|u(\sigma)\|_{B^{\frac52}_{2,1}}\exp\left(C\int_\sigma^\infty\|u(t)\|_{\dot B^{\frac52}_{2,1}}dt\right).$$

**Remark 1.** We note that the smallness assumption is made only on the homogeneous norm of the initial data.

Our second main theorem below is concerned with the global stability of the solution of $(\mathrm{SNS})_\alpha$ in the case $\alpha\ge\frac12$. For the stability of the usual Navier–Stokes equations, Beirão da Veiga–Secchi [1] and Wiegner [25] obtained $L^p$-stability with $p>3$ near the $L^\infty(0,\infty;L^{p+2})$-solution. Ponce–Racke–Sideris–Titi [23] proved the $H^1$-stability of mildly decaying global strong solutions of the Navier–Stokes equations.

**Theorem 2.** *Let $\alpha \in [\frac{1}{2}, \frac{5}{4})$ be given. Assume that $u^1$ is a solution of the IVP* $(\mathrm{SNS})_\alpha$ *satisfying $u^1 \in C([0,\infty); B_{2,1}^{\frac{5}{2}-2\alpha}) \cap L^1(0,\infty; \dot{B}_{2,1}^{\frac{5}{2}})$. Then there exists a positive constant $\epsilon_0 = \epsilon_0(\|u_0^1\|_{B_{2,1}^{2-2\alpha}}, \|u^1\|_{L^1(0,\infty;\dot{B}_{2,1}^{\frac{5}{2}})})$ such that if $\|u_0^1 - u_0^2\|_{\dot{B}_{2,1}^{\frac{5}{2}-2\alpha}} < \epsilon_0$, there exists a unique global solution $u^2 \in C([0,\infty); B_{2,1}^{\frac{5}{2}-2\alpha}) \cap L^1(0,\infty; \dot{B}_{2,1}^{\frac{5}{2}}) \cap C((0,\infty); B_{2,1}^{2})$ of* $(\mathrm{SNS})_\alpha$ *with initial data $u_0^2 \in B_{2,1}^{\frac{5}{2}-2\alpha}$.*

**Remark 2.** Theorem 2 could be viewed as a generalization of Theorem 1. Since $H^s$ $(s > \frac{1}{2})$ is contained in $B_{2,1}^{\frac{1}{2}}$, our result in the case $\alpha = 1$ can be regarded as improvement of the result of the corresponding stability result of the Navier–Stokes equations in [23].

For the usual Navier–Stokes equations $(\mathrm{SNS})_1$, Chemin [15] proved local in time existence in some critical Besov spaces and Cannone–Planchon [8] proved the global existence in some critical Besov spaces if the initial data has a small Besov norm. Cannone–Planchon [9] also derived various estimates for strong solutions in $C([0,T]; L^3(\mathbb{R}^3))$ to the 3-dimensional incompressible Navier–Stokes equations. Using the similar method originated from Fujita–Kato [19] and Kato [20], we can improve parts of Theorem 1 in the case $\frac{1}{2} < \alpha \le \frac{5}{4}$ as follows.

**Theorem 3.** *Let $\alpha \in (\frac{1}{2}, \frac{5}{4}]$ be given. Suppose $1 \le p < \frac{3}{2\alpha-1}$. There exists a constant $\epsilon > 0$ such that for any $u_0 \in \dot{B}_{p,\infty}^{\frac{3}{p}+1-2\alpha}$ and $\|u_0\|_{\dot{B}_{p,\infty}^{\frac{3}{p}+1-2\alpha}} < \epsilon$, the IVP* $(\mathrm{SNS})_\alpha$ *has a global solution $u \in C([0,\infty); \dot{B}_{p,\infty}^{\frac{3}{p}+1-2\alpha})$.*

## 2. Littlewood–Paley decomposition

We first set our notation, and recall definitions of the Besov spaces. We follow [24]. Let $\mathcal{S}$ be the Schwartz class of rapidly decreasing functions. Given $f \in \mathcal{S}$, its Fourier transform $\mathcal{F}(f) = \hat{f}$ is defined by

$$\hat{f}(\xi) = \frac{1}{(2\pi)^{n/2}} \int e^{-ix\cdot\xi} f(x)dx.$$

We consider $\varphi \in \mathcal{S}$ satisfying $\mathrm{Supp}\,\hat{\varphi} \subset \{\xi \in \mathbb{R}^n \mid \frac{1}{2} \le |\xi| \le 2\}$, and $\hat{\varphi}(\xi) > 0$ if $\frac{1}{2} < |\xi| < 2$. Setting $\hat{\varphi}_j = \hat{\varphi}(2^{-j}\xi)$ (in other words, $\varphi_j(x) = 2^{jn}\varphi(2^j x)$), we can adjust the normalization constant in front of $\hat{\varphi}$ so that

$$\sum_{j\in\mathbb{Z}} \hat{\varphi}_j(\xi) = 1 \quad \forall \xi \in \mathbb{R}^n \setminus \{0\}.$$

Given $k \in \mathbb{Z}$, we define the function $S_k \in \mathcal{S}$ by its Fourier transform

$$\hat{S}_k(\xi) = 1 - \sum_{j\ge k+1} \hat{\varphi}_j(\xi).$$

We observe

$$\text{Supp }\hat{\varphi}_j \cap \text{ Supp }\hat{\varphi}_{j'} = \emptyset \text{ if } |j-j'| \geq 2.$$

Let $s \in \mathbb{R}$, $p,q \in [0,\infty]$. Given $f \in \mathcal{S}'$, we denote $\Delta_j f = \varphi_j * f$. Then the homogeneous Besov semi-norm $\|f\|_{\dot{B}^s_{p,q}}$ is defined by

$$\|f\|_{\dot{B}^s_{p,q}} = \begin{cases} \left[\sum_{-\infty}^{\infty} 2^{jqs}\|\varphi_j * f\|^q_{L^p}\right]^{\frac{1}{q}} & \text{if } q \in [1,\infty) \\ \sup_j \left[2^{js}\|\varphi_j * f\|_{L^p}\right] & \text{if } q = \infty. \end{cases}$$

The homogeneous Besov space $\dot{B}^s_{p,q}$ is a quasi-normed space with the quasi-norm given by $\|\cdot\|_{\dot{B}^s_{p,q}}$. For $s>0$ we define the inhomogeneous Besov space norm $\|f\|_{B^s_{p,q}}$ of $f \in \mathcal{S}'$ as $\|f\|_{B^s_{p,q}} = \|f\|_{L^p} + \|f\|_{\dot{B}^s_{p,q}}$. Let us now state some basic properties for the Besov spaces.

**Proposition 1.** (i) *Bernstein's Lemma : Assume that $f \in L^p$, $1 \leq p \leq \infty$, and supp $\hat{f} \subset \{2^{j-2} \leq |\xi| < 2^j\}$. Then there exists a constant $C_k$ such that the following inequality holds:*

$$C_k^{-1}2^{jk}\|f\|_{L^p} \leq \|D^k f\|_{L^p} \leq C_k 2^{jk}\|f\|_{L^p}.$$

(ii) *We have the equivalence of norms*

$$\|D^k f\|_{\dot{B}^s_{p,q}} \sim \|f\|_{\dot{B}^{s+k}_{p,q}}.$$

(iii) *Let $s>0$, $q \in [1,\infty]$, then there exists a constant $C$ such that the inequality*

$$\|fg\|_{\dot{B}^s_{p,q}} \leq C\left(\|f\|_{L^{p_1}}\|g\|_{\dot{B}^s_{p_2,q}} + \|g\|_{L^{r_1}}\|f\|_{\dot{B}^s_{r_2,q}}\right),$$

*holds for homogeneous Besov spaces, where $p_1$, $r_1 \in [1,\infty]$ such that $\frac{1}{p} = \frac{1}{p_1} + \frac{1}{p_2} = \frac{1}{r_1} + \frac{1}{r_2}$.*
*Let $s_1$, $s_2 \leq \frac{N}{p}$ such that $s_1+s_2>0$, $f \in \dot{B}^{s_1}_{p,1}$ and $g \in \dot{B}^{s_2}_{p,1}$. Then $fg \in \dot{B}^{s_1+s_2-\frac{N}{p}}_{p,1}$ and*

$$\|fg\|_{\dot{B}^{s_1+s_2-\frac{N}{p}}_{p,1}} \leq C\|f\|_{\dot{B}^{s_1}_{p,1}}\|g\|_{\dot{B}^{s_2}_{p,1}}.$$

(iv) *If $s \in (-\frac{N}{p}-1, \frac{N}{p}]$, then we have*

$$\|[u,\Delta_q]w\|_{L^p} \leq c_q 2^{-q(s+1)}\|u\|_{\dot{B}^{\frac{N}{p}+1}_{p,1}}\|w\|_{\dot{B}^s_{p,1}}$$

*with $\sum_{q\in\mathbb{Z}} c_q \leq 1$. In the above, we denote*

$$[u,\Delta_q]w = u\Delta_q w - \Delta_q(uw).$$

(v) *(Embedding) $\dot{B}^{\frac{N}{p}}_{p,1}(\mathbb{R}^N)$ is an algebra included in the space $C_0$ of continuous functions tending to $0$ at infinity. Let $s \in \mathbb{R}$, $\epsilon > 0$, and suppose $p,q \in [1.\infty]$. Then it holds*

$$\dot{B}^s_{p,1} \hookrightarrow \dot{H}^s_p \hookrightarrow \dot{B}^s_{p,\infty},$$

*and*

$$B^{s+\epsilon}_{p,\infty} \hookrightarrow B^{s}_{p,q}.$$

(vi) *(Interpolation) We have the following interpolation inequalities for* $s_1$, $s_2 \in \mathbb{R}$, $\theta \in [0,1]$*:*

$$\|u\|_{\dot{B}^{\theta s_1+(1-\theta)s_2}_{p,1}} \le C\|u\|^{\theta}_{\dot{B}^{s_1}_{p,1}}\|u\|^{1-\theta}_{\dot{B}^{s_2}_{p,1}},$$

*and*

$$\|u\|_{B^{\theta s_1+(1-\theta)s_2}_{p,1}} \le C\|u\|^{\theta}_{B^{s_1}_{p,1}}\|u\|^{1-\theta}_{B^{s_2}_{p,1}}.$$

*Proof.* The proof of (i)–(vi) is rather standard and we can find the proof in many references. The proof of (i) can be found in [14] Lemma 2.1.1. (ii) is straightforward from (i). The proof of the first part of (iii) can be found in [11] and the second part of (iii) can be found in [13]. We can find the proof of (iv) in [17]. The fact that $\dot{B}^{\frac{N}{p}}_{p,1}$ is embedded in $C_0$ can be found in [2]. We can find the proof of the last part of (v) and (vi) in [24]. □

For simplicity, in the following we denote $\dot{B}^{s}_{p,\infty}$ and $\dot{B}^{\frac{3}{p}+1-2\alpha}_{p,\infty}$ by $\dot{B}^{s}_{p}$ and $\dot{B}_{p}$, respectively. If $(\rho,p,r)\in[1,\infty]$, we denote

$$\|u\|_{\tilde{L}^{\rho}_{T}(\dot{B}^{s}_{p,r})} = \|(2^{qs}\|\Delta_q u\|_{L^{\rho}(0,T;L^{p})})_{q\in\mathbb{Z}}\|_{l^{r}(\mathbb{Z})}.$$

We denote briefly $L^{\infty}(0,\infty;\dot{B}^{s}_{p,r})$ by $L^{\infty}(\dot{B}^{s}_{p,r})$. We denote $(-\Delta)^{\frac{1}{2}}$ by $\Lambda$ for notational simplicity. For the proof of Theorem 3, we need the following extended version of Proposition 2.1 of [15].

**Proposition 2.** *Let* $\alpha \ge 0$ *be given. There exists a constant* $C>0$ *such that*

$$\|e^{-t\Lambda^{2\alpha}}u_0\|_{\tilde{L}^{\rho}_{T}(\dot{B}^{s+\frac{2\alpha}{\rho}}_{p,r})} \le C\|u_0\|_{\dot{B}^{s}_{p,r}}. \tag{1}$$

*If* $u$ *is a solution of*

$$\begin{cases} \partial_t u - \Lambda^{2\alpha}u = f, & \mathbb{R}^3\times\mathbb{R}_+,\\ u(0,x)=0, \end{cases}$$

*then we have*

$$\|u\|_{\tilde{L}^{\rho}_{T}(\dot{B}^{s+2\alpha}_{p})} \le C\|f\|_{\tilde{L}^{\rho}_{T}(\dot{B}^{s}_{p})}, \tag{2}$$

*and*

$$\|u\|_{\tilde{L}^{\rho_1}_{T}(\dot{B}^{s+2\alpha(1+\frac{1}{\rho_1}-\frac{1}{\rho_2})}_{p})} \le C\|f\|_{\tilde{L}^{\rho_2}_{T}(\dot{B}^{s}_{p})}, \tag{3}$$

*where* $\rho_1 \ge \rho_2$.

We provide the proof of Proposition 2 in the appendix.

Taking the divergence operation on the first equation of $(\mathrm{SNS})_\alpha$, we have the formula

$$-\Delta p = \sum_{j,k}\partial_j\partial_k(u^j u^k).$$

This enables us to define the general sub-dissipative Navier–Stokes type equations

$$\begin{cases} \partial_t u - \Lambda^{2\alpha} u = Q(u,u), & \mathbb{R}^3 \times \mathbb{R}_+, \quad 0 \le \alpha < \frac{5}{4}, \\ u(0,x) = u_0, \end{cases}$$

with $Q(u,u) = -\mathrm{div}(u \otimes u) + \sum_{j,k} \nabla \Delta^{-1} \partial_j \partial_k (u^j u^k)$. This general equations of the usual Navier–Stokes equations was studied by Chemin [15]. The following proposition is more or less standard (see [15]).

**Proposition 3.** *Let* $\frac{1}{\rho_1} + \frac{1}{\rho_2} = \frac{1}{\rho}$. *Set* $s_{1,2} = s_1 + s_2 - \frac{3}{p}$. *If* $s_i < \frac{3}{p}$ *and* $s_1 + s_2 > 0$, *then we have*

$$\|Q(u,v)\|_{\tilde{L}^\rho_T(\dot{B}_p^{s_{1,2}-1})} \le C\|u\|_{\tilde{L}^{\rho_1}_T(\dot{B}_p^{s_1})} \|v\|_{\tilde{L}^{\rho_2}_T(\dot{B}_p^{s_2})}. \tag{4}$$

*Proof.* The proof can be found in [13] and [15]. □

## 3. Proof of Theorems

*Proof of Theorem* 1. If we try to follow the classical Fujita–Kato method ([19] and [20]) to prove Theorem 1, we need to prove the continuity of the bilinear operator in $C([0,\infty); B_{2,1}^{\frac{5}{2}-2\alpha})$,

$$B(u,u) = \int_0^t \mathbb{P} e^{-(t-s)(-\Delta)^\alpha} \nabla \cdot (v \otimes v)(s) ds,$$

where $\mathbb{P}$ is a Leray projection operator, and we found there are technical difficulties for such a continuity estimate when $0 \le \alpha \le \frac{1}{2}$. Thus we could not adapt Fujita–Kato's argument to prove Theorem 1.

Step 1. *A priori estimates*

Taking the operator $\Delta_q$ on the first equation of $(\mathrm{SNS})_\alpha$, we have

$$\partial_t \Delta_q u + (u \cdot \nabla)\Delta_q u + \Lambda^{2\alpha} \Delta_q u + \nabla \Delta_q p = [u, \Delta_q] \cdot \nabla u. \tag{5}$$

Multiplying $\Delta_q u$ on the both sides of (5) and integrating over $\mathbb{R}^3$, we obtain

$$\frac{1}{2}\frac{d}{dt}\|\Delta_q u\|_{L^2}^2 + C 2^{2\alpha q}\|\Delta_q u\|_{L^2}^2 \le \|[\Delta_q, u] \cdot \nabla u\|_{L^2} \|\Delta_q u\|_{L^2}. \tag{6}$$

Dividing the both sides of (6) by $\|\Delta_q u\|_{L^2}$ and using the commutator estimates of Proposition 1 (iv), we have

$$\frac{d}{dt}\|\Delta_q u\|_{L^2} + C 2^{2\alpha q}\|\Delta_q u\|_{L^2} \le C c_q 2^{-(\frac{5}{2}-2\alpha)q} \|u\|_{\dot{B}_{2,1}^{\frac{5}{2}}} \|u\|_{\dot{B}_{2,1}^{\frac{5}{2}-2\alpha}}. \tag{7}$$

Multiplying $2^{(\frac{5}{2}-2\alpha)q}$ on both sides of (7) and taking summation over $q \in \mathbb{Z}$, it reduces to

$$\frac{d}{dt}\|u(t)\|_{\dot{B}_{2,1}^{\frac{5}{2}-2\alpha}} + C_1 \|u\|_{\dot{B}_{2,1}^{\frac{5}{2}}} \le C_2 \|u\|_{\dot{B}_{2,1}^{\frac{5}{2}}} \|u\|_{\dot{B}_{2,1}^{\frac{5}{2}-2\alpha}}.$$

By the Gronwall inequality, we obtain that

$$\begin{aligned}&\sup_{0\le t<\infty}\|u(t)\|_{\dot B^{\frac52-2\alpha}_{2,1}}+C_1\int_0^\infty\|u(t)\|_{\dot B^{\frac52}_{2,1}}\,dt\\&\le\|u_0\|_{\dot B^{\frac52-2\alpha}_{2,1}}\exp\left(C_2\int_0^\infty\|u(t)\|_{\dot B^{\frac52}_{2,1}}\,dt\right).\end{aligned}\tag{8}$$

We also note that

$$\sup_{0\le t<\infty}\|u(t)\|_{L^2}\le\|u_0\|_{L^2}.$$

Then we have the following a priori estimates.

$$\begin{aligned}&\sup_{0\le t<\infty}\|u(t)\|_{B^{\frac52-2\alpha}_{2,1}}+C_1\int_0^\infty\|u(t)\|_{\dot B^{\frac52}_{2,1}}\,dt\\&\le\|u_0\|_{B^{\frac52-2\alpha}_{2,1}}\exp\left(C_2\int_0^\infty\|u(t)\|_{\dot B^{\frac52}_{2,1}}\,dt\right).\end{aligned}\tag{9}$$

Step 2. *Iteration and Uniform estimates*

We define the following iteration sequences ( $u^{n+1}$, $p^{n+1}$).

$$(\mathrm{I})\begin{cases}\partial_t u^{n+1}+(u^n\cdot\nabla)u^{n+1}+\Lambda^{2\alpha}u^{n+1}+\nabla p^{n+1}=0, & \mathbb{R}^3\times\mathbb{R}_+,\quad 0\le 2\alpha<\frac52,\\ \operatorname{div}u^{n+1}=0,\\ u^{n+1}(x,0)=u_0^{n+1}(x)=\sum_{q\le n+1}\Delta_q u_0.\end{cases}$$

Setting $u^0=0$, we can find $u^n$ and $p^n$ for all $n$ by solving the linear system. Similarly to a priori estimates, we have

$$\frac{d}{dt}\|u^{n+1}(t)\|_{\dot B^{\frac52-2\alpha}_{2,1}}+C_1\|u^{n+1}\|_{\dot B^{\frac52}_{2,1}}\le C_2\|u^n\|_{\dot B^{\frac52}_{2,1}}\|u^{n+1}\|_{\dot B^{\frac52-2\alpha}_{2,1}}.$$

By the Gronwall inequality, we obtain

$$\begin{aligned}&\sup_{0\le t<\infty}\|u^{n+1}(t)\|_{\dot B^{\frac52-2\alpha}_{2,1}}+C_1\int_0^\infty\|u^{n+1}(t)\|_{\dot B^{\frac52}_{2,1}}\,dt\\&\le\|u_0^{n+1}\|_{\dot B^{\frac52-2\alpha}_{2,1}}\exp\left(C_2\int_0^\infty\|u^n(t)\|_{\dot B^{\frac52}_{2,1}}\,dt\right).\end{aligned}\tag{10}$$

Choose $\epsilon$ so small that $\exp\left(\frac{C_2M\epsilon}{C_1}\right)\le M$ (this is possible if we take $M>1$). Assume that $\|u^n\|_{L^\infty(0,T;\dot B^{\frac52-2\alpha}_{2,1})}+C_1\|u^n\|_{L^1(0,T;\dot B^{\frac52}_{2,1})}\le M\epsilon$. From the estimate of (10), we have

$$\begin{aligned}
&\sup_{0\le t<\infty}\|u^{n+1}(t)\|_{\dot B^{\frac52-2\alpha}_{2,1}}+C_1\int_0^\infty\|u^{n+1}(t)\|_{\dot B^{\frac52}_{2,1}}\,dt\\
&\le\|u_0^{n+1}\|_{\dot B^{\frac52-2\alpha}_{2,1}}\exp\left(C_2\int_0^\infty\|u^n(t)\|_{\dot B^{\frac52}_{2,1}}\,dt\right)\\
&\le\|u_0^{n+1}\|_{\dot B^{\frac52-2\alpha}_{2,1}}\exp\left(\frac{C_2M\epsilon}{C_1}\right)\\
&\le M\epsilon.
\end{aligned}$$

Thus we obtain that if $\|u_0\|_{\dot B^{\frac52-2\alpha}_{2,1}}\le\epsilon$, then for all $n$,

$$\sup_{0\le t<\infty}\|u^{n+1}(t)\|_{\dot B^{\frac52-2\alpha}_{2,1}}+C_1\int_0^\infty\|u^{n+1}(t)\|_{\dot B^{\frac52}_{2,1}}\,dt\le M\epsilon, \tag{11}$$

for some $M>0$.

Step 3. *Equations of differences and Existence of solution.*

Define $\delta u^{n+1}=u^{n+1}-u^n$ and $\delta p^{n+1}=p^{n+1}-p^n$. We obtain the equations of the differences,

$$(\mathrm{I}')\begin{cases}\partial_t\delta u^{n+1}+(u^n\cdot\nabla)\delta u^{n+1}+(\delta u^n\cdot\nabla)u^n\\
\quad+\Lambda^{2\alpha}\delta u^{n+1}+\nabla\delta p^{n+1}=0,\quad\mathbb{R}^3\times\mathbb{R}_+,\quad 0\le 2\alpha<\frac52,\\
\mathrm{div}\ \delta u^{n+1}=0,\\
\delta u^{n+1}(x,0)=\Delta_{n+1}u_0.\end{cases}$$

Similarly to a priori estimates, we have for $\eta$ satisfying $\eta=\max\{0,1-2\alpha\}$,

$$\begin{aligned}
&\frac12\frac{d}{dt}\|\Delta_q\delta u^{n+1}\|_{L^2}^2+C2^{2\alpha q}\|\Delta_q\delta u^{n+1}\|_{L^2}^2\\
&\le Cc_q2^{-q(\frac52-2\alpha-\eta)}\|u^n\|_{\dot B^{\frac52}_{2,1}}\|\delta u^{n+1}\|_{\dot B^{\frac52-2\alpha-\eta}_{2,1}}\|\Delta_q\delta u^{n+1}\|_{L^2}\\
&\quad+C\|\Delta_q((\delta u^n\cdot\nabla)u^n)\|_{L^2}\|\Delta_q\delta u^{n+1}\|_{L^2}.
\end{aligned}$$

Dividing both sides of the above inequality by $\|\Delta_q\delta u^{n+1}\|_{L^2}$, multiplying with $2^{q(\frac52-2\alpha-\eta)}$, and taking summation over $q$, we infer that

$$\begin{aligned}
&\frac{d}{dt}\|\delta u^{n+1}\|_{\dot B^{\frac52-2\alpha-\eta}_{2,1}}+C_3\|\delta u^{n+1}\|_{\dot B^{\frac52-\eta}_{2,1}}\\
&\le C_4\|u^n\|_{\dot B^{\frac52}_{2,1}}\|\delta u^{n+1}\|_{\dot B^{\frac52-2\alpha-\eta}_{2,1}}\\
&\quad+C_5\|\delta u^n\|_{\dot B^{\frac52-2\alpha-\eta}_{2,1}}\|u^n\|_{\dot B^{\frac52}_{2,1}}.
\end{aligned}$$

In the above inequality, we used Proposition 1 (iii), e.g.,

$$\|((\delta u^n)\cdot\nabla)u^n\|_{\dot B^{\frac52-2\alpha-\eta}_{2,1}}\le C\|\delta u^n\|_{\dot B^{\frac52-2\alpha-\eta}_{2,1}}\|\nabla u^n\|_{\dot B^{\frac32}_{2,1}}.$$

Gronwall's inequality gives us that

$$\begin{aligned}
&\sup_{0\le t<\infty}\|\delta u^{n+1}(t)\|_{\dot B_{2,1}^{\frac52-2\alpha-\eta}}+C_3\int_0^\infty\|\delta u^{n+1}(t)\|_{\dot B_{2,1}^{\frac52-\eta}}dt\\
&\le\|\delta u_0^{n+1}\|_{\dot B_{2,1}^{\frac52-2\alpha-\eta}}\exp\left(C_4\int_0^\infty\|u^n(t)\|_{\dot B_{2,1}^{\frac52}}dt\right)\\
&\quad+C_5\sup_{0\le t<\infty}\|\delta u^n(t)\|_{\dot B_{2,1}^{\frac52-2\alpha-\eta}}\int_0^\infty\|u^n\|_{\dot B_{2,1}^{\frac52}}dt\exp\left(C_4\int_0^\infty\|u^n(\tau)\|_{\dot B_{2,1}^{\frac52}}d\tau\right).
\end{aligned}\tag{12}$$

From the uniform estimates (11), we can choose $\epsilon$ so small that

$$\exp\left(C_4\int_0^\infty\|u^n(t)\|_{\dot B_{2,1}^{\frac52}}dt\right)\le 2$$

and

$$C_5\int_0^\infty\|u^n(t)\|_{\dot B_{2,1}^{\frac52}}dt<\frac18,$$

for all $n$. Then we have

$$\begin{aligned}
&\sup_{0\le t<\infty}\|\delta u^{n+1}(t)\|_{\dot B_{2,1}^{\frac52-2\alpha-\eta}}+C_3\int_0^\infty\|\delta u^{n+1}(t)\|_{\dot B_{2,1}^{\frac52-\eta}}dt\\
&\le 2\|\delta u_0^{n+1}\|_{\dot B_{2,1}^{\frac52-2\alpha-\eta}}+\frac14\sup_{0\le t<\infty}\|\delta u^n(t)\|_{\dot B_{2,1}^{\frac52-2\alpha-\eta}}.
\end{aligned}\tag{13}$$

By iterating, we have for any positive integer $N>0$,

$$\begin{aligned}
&\sum_{n=1}^N\sup_{0\le t<\infty}\|\delta u^{n+1}(t)\|_{\dot B_{2,1}^{\frac52-2\alpha-\eta}}+C_3\sum_{n=1}^N\int_0^\infty\|\delta u^{n+1}(t)\|_{\dot B_{2,1}^{\frac52-\eta}}dt\\
&\le 2\sum_{n=1}^N\|\delta u_0^{n+1}\|_{\dot B_{2,1}^{\frac52-2\alpha-\eta}}+\frac14\sum_{n=2}^N\sup_{0\le t<\infty}\|\delta u^n(t)\|_{\dot B_{2,1}^{\frac52-2\alpha-\eta}}\\
&\le 2C\sum_{n=1}^N2^{(n+1)(\frac52-2\alpha-\eta)}\|\Delta_{n+1}u_0\|_{L^2}+\frac14\sum_{n=2}^N\sup_{0\le t<\infty}\|\delta u^n(t)\|_{\dot B_{2,1}^{\frac52-2\alpha-\eta}}\\
&\le 2C\|u_0\|_{\dot B_{2,1}^{\frac52-2\alpha-\eta}}+\frac14\sum_{n=2}^N\sup_{0\le t<\infty}\|\delta u^n(t)\|_{\dot B_{2,1}^{\frac52-2\alpha-\eta}}\\
&\le 2C\|u_0\|_{\dot B_{2,1}^{\frac52-2\alpha-\eta}}+\dots+\frac{2C}{4^{N-1}}\|u_0\|_{\dot B_{2,1}^{\frac52-2\alpha-\eta}}\\
&\le\frac{8C}{3}\|u_0\|_{\dot B_{2,1}^{\frac52-2\alpha-\eta}}.
\end{aligned}$$

Multiplying both sides of the first equation of (I′) by $\delta u^{n+1}$ and integrating over $\mathbb{R}^3$, we have

$$\sup_{0\le t<\infty}\|\delta u^{n+1}(t)\|_{L^2}\le\|\delta u_0^{n+1}\|_{L^2}+C_6\sup_{0\le t<\infty}\|\delta u^n(t)\|_{L^2}\int_0^\infty\|u^n(t)\|_{\dot B_{2,1}^{\frac52}}dt.$$

If we choose $\epsilon$ so small that $C_6\int_0^\infty \|u^n(t)\|_{\dot B^{\frac52}_{2,1}}\,dt<\frac14$, then we obtain

$$\sup_{0\le t<\infty}\|\delta u^{n+1}(t)\|_{L^2}\le \|\delta u_0^{n+1}\|_{L^2}+\frac14\sup_{0\le t<\infty}\|\delta u^n(t)\|_{L^2}. \tag{14}$$

By iterating, we obtain that $u^n$ converges to $u$ in $L^\infty(0,\infty;B^{\frac52-2\alpha-\eta}_{2,1})\cap L^1(0,\infty;\dot B^{\frac52-\eta}_{2,1})$. From the uniform estimates we have $u\in L^\infty(0,\infty;B^{\frac52-2\alpha}_{2,1})\cap L^1(0,\infty;\dot B^{\frac52}_{2,1})$. Uniqueness can be proved similarly. $u^{n+1}$ satisfies

$$\partial_t u^{n+1}=-(u^n\cdot\nabla)u^{n+1}-\Lambda^{2\alpha}u^{n+1}-\nabla p^{n+1}.$$

Clearly we have $(u^n\cdot\nabla)u^{n+1}$, $\Lambda^{2\alpha}u^{n+1}\in L^1(0,\infty;B^{\frac52-2\alpha}_{2,1})$. Since $p^{n+1}$ satisfies

$$-\mathrm{div}\nabla p^{n+1}=\mathrm{div}((u^n\cdot\nabla)u^{n+1}),$$

we have $\nabla p^{n+1}\in L^1(0,\infty;B^{\frac52-2\alpha}_{2,1})$ from the Calderon–Zygmund estimate [18]. Thus we readily have that $u^{n+1}$ is continuous with values in $B^{\frac52-2\alpha}_{2,1}$. By a $3\epsilon$ argument, $u$ is continuous with values in $B^{\frac52-2\alpha-\eta}_{2,1}$. By the interpolation, we have $C([0,\infty);B^\beta_{2,1})$ with $\beta=\begin{cases}\frac52-2\alpha, & \text{if } \frac12\le\alpha\le\frac54\\ \frac52-2\alpha-\delta_1, & \text{if } 0\le\alpha<\frac12,\end{cases}$ for $\delta_1>0$. We still have to prove that $u$ also belongs to $L^\infty(\sigma,\infty;B^{\frac52}_{2,1})\cap L^1(\sigma,\infty;\dot B^{\frac52+2\alpha}_{2,1})\cap C((0,\infty);B^\gamma_{2,1})$, where $\sigma>0$ and $\gamma=\begin{cases}\frac52-\delta_2, & \text{if } 0\le\alpha<\frac12,\\ \frac52, & \text{if } \frac12\le\alpha<\frac54,\end{cases}$ for any $\delta_2>0$. We provide only a priori estimates. Since we already know that $u$ belongs to $L^1(0,\infty;\dot B^{\frac52}_{2,1})$, we can choose $\sigma>0$ such that $u(\sigma)\in B^{\frac52}_{2,1}$. Similarly to step 1, we have

$$\frac{d}{dt}\|\Delta_q u\|_{L^2}+C2^{2\alpha q}\|\Delta_q u\|_{L^2}\le Cc_q 2^{-\frac52 q}\|u\|_{\dot B^{\frac52}_{2,1}}\|u\|_{\dot B^{\frac52}_{2,1}}.$$

Multiplying $2^{\frac52 q}$, and taking the summation over $q$, we have

$$\frac{d}{dt}\|u\|_{\dot B^{\frac52}_{2,1}}+C\|u\|_{\dot B^{\frac52+2\alpha}_{2,1}}\le C\|u\|^2_{\dot B^{\frac52}_{2,1}}.$$

Using Gronwall's inequality, we have

$$\sup_{\sigma\le t<\infty}\|u(t)\|_{B^{\frac52}_{2,1}}+C\int_\sigma^\infty\|u(t)\|_{\dot B^{\frac52+2\alpha}_{2,1}}dt$$
$$\le\|u(\sigma)\|_{B^{\frac52}_{2,1}}\exp\left(C\int_\sigma^\infty\|u(t)\|_{\dot B^{\frac52}_{2,1}}dt\right).$$

Thus $u$ belongs to $L^\infty(\sigma,\infty;B^{\frac{5}{2}}_{2,1})\cap L^1(\sigma,\infty;\dot{B}^{\frac{5}{2}+2\alpha}_{2,1})$. By the interpolation theorem, we have $u\in C([\sigma,\infty);B^{\frac{5}{2}-\delta_2}_{2,1})$ for any $\delta_2>0$ if $0\le\alpha<\frac{1}{2}$. If $\alpha\ge\frac{1}{2}$, $u$ satisfies $\partial_t u=-(u\cdot\nabla)u-\Lambda^{2\alpha}u-\nabla p$. Since $(u\cdot\nabla)\theta$, $\Lambda^{2\alpha}u$, $\nabla p\in L^1(\sigma,\infty;\dot{B}^{\frac{5}{2}}_{2,1})$, $u$ belongs to $C([\sigma,\infty);B^{\frac{5}{2}}_{2,1})$. This completes the proof of Theorem 1. □

*Proof of Theorem* 2. We derive a differential inequality for $\|u^1(t)-u^2(t)\|_{\dot{B}^{\frac{5}{2}-2\alpha}_{2,1}}$. The difference $U=u^1-u^2$ satisfies the system

$$
(\mathrm{A})\begin{cases}\partial_t U+(U\cdot\nabla)u^1-(U\cdot\nabla)U+(u^1\cdot\nabla)U\\ \qquad +\Lambda^{2\alpha}U+\nabla P=0, \quad \mathbb{R}^3\times\mathbb{R}_+, \quad 1\le 2\alpha<\frac{5}{2},\\ \operatorname{div} U=0,\\ U(0,x)=u_0^1-u_0^2,\end{cases}
$$

where $P=p^1-p^2$.

Step 1. *Iteration and Uniform estimates*

We define the following iterating sequences $(U^{n+1},P^{n+1})$.

$$
(\mathrm{II})\begin{cases}\partial_t U^{n+1}+(U^{n+1}\cdot\nabla)u^1-(U^n\cdot\nabla)U^{n+1}+(u^1\cdot\nabla)U^{n+1}\\ \qquad +\Lambda^{2\alpha}U^{n+1}+\nabla P^{n+1}=0, \quad \mathbb{R}^3\times\mathbb{R}_+, \quad \frac{1}{2}\le\alpha<\frac{5}{4},\\ \operatorname{div} U^{n+1}=0,\\ U(0,x)=\sum_{q\le n+1}(\Delta_q u_0^1-\Delta_q u_0^2).\end{cases}
$$

Since $u^1$ is given, we can find $(U^n,P^n)$ for all $n$ by setting $U^0=0$. Taking the $\Delta_q$ operator on both sides of the first equation of (II), we have

$$
\begin{aligned}&\partial_t\Delta_q U^{n+1}+\Lambda^{2\alpha}\Delta_q U^{n+1}-(U^n\cdot\nabla)\Delta_q U^{n+1}+(u^1\cdot\nabla)\Delta_q U^{n+1}+\nabla\Delta_q P^{n+1}\\ &=-[U^n,\Delta_q]\cdot\nabla U^{n+1}+[u^1,\Delta_q]\cdot\nabla U^{n+1}-\Delta_q((U^{n+1}\cdot\nabla)u^1). \qquad (15)\end{aligned}
$$

Multiplying with $\Delta_q\delta U^{n+1}$ on both sides of (15), integrating over $\mathbb{R}^3$, and dividing both sides by $\|\Delta_q U^{n+1}\|_{L^2}$, we have

$$
\begin{aligned}&\frac{d}{dt}\|\Delta_q U^{n+1}\|_{L^2}+C2^{2\alpha q}\|\Delta_q U^{n+1}\|_{L^2}\\ &\le\|[U^n,\Delta_q]\cdot\nabla U^{n+1}\|_{L^2}+\|[u^1,\Delta_q]\cdot\nabla U^{n+1}\|_{L^2}+\|\Delta_q((U^{n+1}\cdot\nabla)u^1)\|_{L^2}. \qquad (16)\end{aligned}
$$

Using commutator estimates, multiplying with $2^{(\frac{5}{2}-2\alpha)q}$ on both sides of (16), and taking summation over $q$, we obtain that

$$
\begin{aligned}&\frac{d}{dt}\|U^{n+1}\|_{\dot{B}^{\frac{5}{2}-2\alpha}_{2,1}}+C_7\|U^{n+1}\|_{\dot{B}^{\frac{5}{2}}_{2,1}}\\ &\le C\|U^n\|_{\dot{B}^{\frac{5}{2}}_{2,1}}\|U^{n+1}\|_{\dot{B}^{\frac{5}{2}-2\alpha}_{2,1}}+C\|u^1\|_{\dot{B}^{\frac{5}{2}}_{2,1}}\|U^{n+1}\|_{\dot{B}^{\frac{5}{2}-2\alpha}_{2,1}}+C\|(U^{n+1}\cdot\nabla)u^1\|_{\dot{B}^{\frac{5}{2}-2\alpha}_{2,1}}\\ &\le C_8\|U^n\|_{\dot{B}^{\frac{5}{2}}_{2,1}}\|U^{n+1}\|_{\dot{B}^{\frac{5}{2}-2\alpha}_{2,1}}+C_9\|u^1\|_{\dot{B}^{\frac{5}{2}}_{2,1}}\|U^{n+1}\|_{\dot{B}^{\frac{5}{2}-2\alpha}_{2,1}}. \qquad (17)\end{aligned}
$$

In the above last inequality, we used Proposition 1 (iii).

We show that if $\epsilon_0$ is chosen to be so small that

$$\epsilon_0 \exp\left( C_9 \int_0^\infty \|u^1(t)\|_{\dot{B}^{\frac{5}{2}}_{2,1}} dt \right) < \frac{C_7}{2C_8}, \tag{18}$$

then it follows

$$(\mathrm{Q_n}) \qquad \sup_{0\le t<\infty} \|U^n(t)\|_{\dot{B}^{\frac{5}{2}-2\alpha}_{2,1}} + \frac{C_7}{2}\int_0^\infty \|U^n\|_{\dot{B}^{\frac{5}{2}}_{2,1}} dt \le \epsilon_0 \, \exp\left( C_9 \int_0^\infty \|u^1\|_{\dot{B}^{\frac{5}{2}}_{2,1}} dt \right).$$

Suppose that $(\mathrm{Q_n})$ is satisfied and let us prove that $(\mathrm{Q_{n+1}})$ is also true. From (17) we obtain

$$\frac{d}{dt}\|U^{n+1}\|_{\dot{B}^{\frac{5}{2}-2\alpha}_{2,1}} + (C_7 - C_8 \sup_{0\le t<\infty} \|U^n(t)\|_{\dot{B}^{\frac{5}{2}-2\alpha}_{2,1}})\|U^{n+1}\|_{\dot{B}^{\frac{5}{2}}_{2,1}} \le C_9\|u^1\|_{\dot{B}^{\frac{5}{2}}_{2,1}} \|U^{n+1}\|_{\dot{B}^{\frac{5}{2}-2\alpha}_{2,1}}. \tag{19}$$

Since $\epsilon_0$ was chosen to satisfy (18), we have

$$\frac{d}{dt}\|U^{n+1}\|_{\dot{B}^{\frac{5}{2}-2\alpha}_{2,1}} + \frac{C_7}{2}\|U^{n+1}\|_{\dot{B}^{\frac{5}{2}}_{2,1}} \le C_9\|u^1\|_{\dot{B}^{\frac{5}{2}}_{2,1}} \|U^{n+1}\|_{\dot{B}^{\frac{5}{2}-2\alpha}_{2,1}}.$$

Using Gronwall's inequality, we have

$$\sup_{0\le t<\infty} \|U^{n+1}(t)\|_{\dot{B}^{\frac{5}{2}-2\alpha}_{2,1}} + \frac{C_7}{2}\int_0^\infty \|U^{n+1}(t)\|_{\dot{B}^{\frac{5}{2}}_{2,1}} dt \le \|U_0^{n+1}\|_{\dot{B}^{\frac{5}{2}-2\alpha}_{2,1}} \exp\left( C_9 \int_0^\infty \|u^1\|_{\dot{B}^{\frac{5}{2}}_{2,1}} dt \right). \tag{20}$$

Since we have $\|U_0^{n+1}\|_{\dot{B}^{\frac{5}{2}-2\alpha}_{2,1}} \le \|U_0\|_{\dot{B}^{\frac{5}{2}-2\alpha}_{2,1}} \le \epsilon_0$, $(\mathrm{Q_{n+1}})$ is true. Multiplying with $U^{n+1}$ on both sides of the first equation of (II) and integrating over $\mathbb{R}^3$, we infer that

$$\frac{d}{dt}\|U^{n+1}\|^2_{L^2} \le \|U^{n+1}\|^2_{L^2}\|\nabla u^1\|_{\dot{B}^{\frac{3}{2}}_{2,1}}.$$

Using Gronwall's inequality, we have

$$\sup_{0\le t<\infty} \|U^{n+1}\|_{L^2} \le C\|U_0^{n+1}\|_{L^2} \exp\left( C \int_0^\infty \|u^1\|_{\dot{B}^{\frac{5}{2}}_{2,1}} dt \right). \tag{21}$$

From (20) and (21), we obtain

$$\sup_{0\le t<\infty} \|U^{n+1}(t)\|_{B^{\frac{5}{2}-2\alpha}_{2,1}} + \frac{C_7}{2}\int_0^\infty \|U^{n+1}(t)\|_{\dot{B}^{\frac{5}{2}}_{2,1}} dt \le C\|U_0^{n+1}\|_{B^{\frac{5}{2}-2\alpha}_{2,1}} \exp\left( C \int_0^\infty \|u^1\|_{\dot{B}^{\frac{5}{2}}_{2,1}} dt \right). \tag{22}$$

Step 2. *Equations of Difference and existence*

Setting $\delta U^{n+1} = U^{n+1} - U^n$ and $\delta P^{n+1} = P^{n+1} - P^n$, we have the following equations of differences.

$$(\mathrm{II}')\begin{cases} \partial_t \delta U^{n+1} - (U^n\cdot\nabla)\delta U^{n+1} - (\delta U^n\cdot\nabla)U^n \\ \qquad +\Lambda^{2\alpha}\delta U^{n+1} + (\delta U^{n+1}\cdot\nabla)u^1 + (u^1\cdot\nabla)\delta U^{n+1} + \nabla\delta P^{n+1} = 0, \\ \operatorname{div}\,\delta U^{n+1} = 0, \\ \delta U(0,x) = (\Delta_{n+1}u_0^1 - \Delta_{n+1}u_0^2). \end{cases}$$

Taking $\Delta_q$ on both sides of the first equation of (II′), we infer that

$$\begin{aligned} &\partial_t\Delta_q\delta U^{n+1} - (U^n\cdot\nabla)\Delta_q\delta U^{n+1} + (u^1\cdot\nabla)\Delta_q\delta U^{n+1} + \Lambda^{2\alpha}\Delta_q\delta U^{n+1} + \nabla\Delta_q\delta P^{n+1} \\ &= -[U^n,\Delta_q]\cdot\nabla\delta U^{n+1} + [u^1,\Delta_q]\cdot\delta U^{n+1} \\ &\quad -\Delta_q((\delta U^{n+1}\cdot\nabla)u^1) + \Delta_q((\delta U^n\cdot\nabla)U^n). \end{aligned} \tag{23}$$

Multiplying with $\Delta_q\delta U^{n+1}$ on both sides of (23) and integrating over $\mathbb{R}^3$, we have

$$\begin{aligned} &\frac{1}{2}\frac{d}{dt}\|\Delta_q\delta U^{n+1}\|_{L^2}^2 + C2^{2\alpha q}\|\Delta_q\delta U^{n+1}\|_{L^2}^2 \\ &\le \|[U^n,\Delta_q]\cdot\nabla\delta U^{n+1}\|_{L^2}\|\Delta_q\delta U^{n+1}\|_{L^2} + \|[u^1,\Delta_q]\cdot\nabla\delta U^{n+1}\|_{L^2}\|\Delta_q\delta U^{n+1}\|_{L^2} \\ &+\|\Delta_q((\delta U^n\cdot\nabla)U^n)\|_{L^2}\|\Delta_q\delta U^{n+1}\|_{L^2} + \|\Delta_q((\delta U^{n+1}\cdot\nabla)u^1)\|_{L^2}\|\Delta_q\delta U^{n+1}\|_{L^2}. \end{aligned} \tag{24}$$

Dividing both sides of (24) by $\|\Delta_q\delta U^{n+1}\|_{L^2}$, multiplying with $2^{(\frac{5}{2}-2\alpha)q}$ on both sides, taking summation over $q$, and using (iii) and (iv) of Proposition 1, we obtain

$$\begin{aligned} &\frac{d}{dt}\|\delta U^{n+1}\|_{\dot B^{\frac52-2\alpha}_{2,1}} + C_{10}\|\delta U^{n+1}\|_{\dot B^{\frac52}_{2,1}} \\ &\le C\|U^n\|_{\dot B^{\frac52}_{2,1}}\|\delta U^{n+1}\|_{\dot B^{\frac52-2\alpha}_{2,1}} + C\|u^1\|_{\dot B^{\frac52}_{2,1}}\|\delta U^{n+1}\|_{\dot B^{\frac52-2\alpha}_{2,1}} \\ &\quad +C\|\delta U^n\|_{\dot B^{\frac52-2\alpha}_{2,1}}\|\nabla U^n\|_{\dot B^{\frac32}_{2,1}} + C\|\delta U^{n+1}\|_{\dot B^{\frac52-2\alpha}_{2,1}}\|\nabla u^1\|_{\dot B^{\frac32}_{2,1}} \\ &\le C_{11}\|U^n\|_{\dot B^{\frac52}_{2,1}}\|\delta U^{n+1}\|_{\dot B^{\frac52-2\alpha}_{2,1}} + C_{12}\|u^1\|_{\dot B^{\frac52}_{2,1}}\|\delta U^{n+1}\|_{\dot B^{\frac52-2\alpha}_{2,1}} \\ &\quad +C_{13}\|U^n\|_{\dot B^{\frac52}_{2,1}}\|\delta U^n\|_{\dot B^{\frac52-2\alpha}_{2,1}}. \end{aligned}$$

Using Gronwall's inequality, we have

$$\begin{aligned} &\sup_{0\le t<\infty}\|\delta U^{n+1}(t)\|_{\dot B^{\frac52-2\alpha}_{2,1}} + C_{10}\int_0^\infty \|\delta U^{n+1}(t)\|_{\dot B^{\frac52}_{2,1}}\,dt \\ &\le \|\delta U_0^{n+1}\|_{\dot B^{\frac52-2\alpha}_{2,1}}\exp\left(\int_0^\infty C_{11}\|U^n\|_{\dot B^{\frac52}_{2,1}} + C_{12}\|u^1\|_{\dot B^{\frac52}_{2,1}}\,dt\right) \\ &\quad +C_{13}\sup_{0\le t<\infty}\|\delta U^n(t)\|_{\dot B^{\frac52-2\alpha}_{2,1}}\int_0^\infty\|U^n\|_{\dot B^{\frac52-2\alpha}_{2,1}}dt \\ &\quad \times\exp\left(\int_0^\infty C_{11}\|U^n\|_{\dot B^{\frac52}_{2,1}} + C_{12}\|u^1\|_{\dot B^{\frac52}_{2,1}}\,dt\right). \end{aligned} \tag{25}$$

From (20) we have

$$\int_0^\infty \|U^n(t)\|_{\dot{B}^{\frac{5}{2}}_{2,1}} dt \le \frac{2\epsilon_0}{C_7} \exp\left( C_9 \int_0^\infty \|u^1\|_{\dot{B}^{\frac{5}{2}}_{2,1}} dt \right). \tag{26}$$

Substituting (26) into (25), we obtain

$$\begin{aligned}
&\sup_{0\le t<\infty} \|\delta U^{n+1}(t)\|_{\dot{B}^{\frac{5}{2}-2\alpha}_{2,1}} + C_{10} \int_0^\infty \|\delta U^{n+1}(t)\|_{\dot{B}^{\frac{5}{2}}_{2,1}} dt \\
&\le \left( \|\delta U_0^{n+1}\|_{\dot{B}^{\frac{5}{2}-2\alpha}_{2,1}} + C_{13} \sup_{0\le t<\infty} \|\delta U^n(t)\|_{\dot{B}^{\frac{5}{2}-2\alpha}_{2,1}} \int_0^\infty \|U^n\|_{\dot{B}^{\frac{5}{2}-2\alpha}_{2,1}} dt \right) \\
&\quad \times \exp\left( \frac{2C_{11}\epsilon_0}{C_7} \exp\left( C_9 \|u^1\|_{L^1(0,\infty;\dot{B}^{\frac{5}{2}}_{2,1})} \right) + C_{12} \|u^1\|_{L^1(0,\infty;\dot{B}^{\frac{5}{2}}_{2,1})} \right).
\end{aligned}$$

Setting

$$M_1 = \exp\left( \frac{2C_{11}\epsilon_0}{C_7} \exp\left( C_9 \|u^1\|_{L^1(0,\infty;\dot{B}^{\frac{5}{2}}_{2,1})} \right) + C_{12} \|u^1\|_{L^1(0,\infty;\dot{B}^{\frac{5}{2}}_{2,1})} \right),$$

we choose $\epsilon_0$ so small that

$$\frac{2\epsilon_0}{C_7} \exp\left( C_9 \int_0^\infty \|u^1\|_{\dot{B}^{\frac{5}{2}}_{2,1}} dt \right) \le \frac{1}{4C_{13}M_1}.$$

Then we have

$$\begin{aligned}
&\sup_{0\le t<\infty} \|\delta U^{n+1}(t)\|_{\dot{B}^{\frac{5}{2}-2\alpha}_{2,1}} + C_{10} \int_0^\infty \|\delta U^{n+1}(t)\|_{\dot{B}^{\frac{5}{2}}_{2,1}} dt \\
&\le M_1 \|\delta U_0^{n+1}\|_{\dot{B}^{\frac{5}{2}-2\alpha}_{2,1}} + \frac{1}{4} \sup_{0\le t<\infty} \|\delta U^n(t)\|_{\dot{B}^{\frac{5}{2}-2\alpha}_{2,1}}.
\end{aligned}$$

Multiplying with $\delta U^{n+1}$ on both sides of (IV$'$) and integrating over $\mathbb{R}^3$, we have

$$\begin{aligned}
&\frac{1}{2}\frac{d}{dt} \|\delta U^{n+1}\|_{L^2}^2 + \|\Lambda^\alpha \delta U^{n+1}\|_{L^2}^2 \\
&\le \|\nabla U^n\|_{L^\infty} \|\delta U^n\|_{L^2} \|\delta U^{n+1}\|_{L^2} + \|\nabla u^1\|_{L^\infty} \|\delta U^{n+1}\|_{L^2} \|\delta U^{n+1}\|_{L^2} \\
&\le C_{14} \|U^n\|_{\dot{B}^{\frac{5}{2}}_{2,1}} \|\delta U^n\|_{L^2} \|\delta U^{n+1}\|_{L^2} + C_{15} \|u^1\|_{\dot{B}^{\frac{5}{2}}_{2,1}} \|\delta U^{n+1}\|_{L^2}^2.
\end{aligned} \tag{27}$$

Dividing both sides of (27) by $\|\delta U^{n+1}\|_{L^2}$ and using Gronwall's inequality, we obtain

$$\begin{aligned}
&\sup_{0\le t<\infty} \|\delta U^{n+1}\|_{L^2} \le \|\delta U_0^{n+1}\|_{L^2} \exp\left( C_{15} \int_0^\infty \|u^1\|_{\dot{B}^{\frac{5}{2}}_{2,1}} dt \right) \\
&+ C_{14} \sup_{0\le t<\infty} \|\delta U^n\|_{L^2} \exp\left( C_{15} \int_0^\infty \|u^1\|_{\dot{B}^{\frac{5}{2}}_{2,1}} dt \right) \int_0^\infty \|U^n\|_{\dot{B}^{\frac{5}{2}}_{2,1}} dt.
\end{aligned}$$

Setting $M_2 = \exp\left( C_{15} \int_0^\infty \|u^1\|_{\dot{B}^{\frac{5}{2}}_{2,1}} dt \right)$, choose $\epsilon_0$ so small that

$$\frac{2\epsilon_0}{C_7} \exp\left( C_9 \int_0^\infty \|u^1\|_{\dot{B}^{\frac{5}{2}}_{2,1}} dt \right) \le \frac{1}{4C_{14}M_2}.$$

Thus we have

$$\sup_{0\le t<\infty}\|\delta U^{n+1}(t)\|_{L^2}\le M_2\|\delta U_0^{n+1}\|_{L^2}+\frac{1}{4}\sup_{0\le t<\infty}\|\delta U^n(t)\|_{L^2}.$$

By the iteration, $U^n$ converges to $U$ in $L^\infty(0,\infty;B_{2,1}^{\frac52-2\alpha})\cap L^1(0,\infty;\dot B_{2,1}^{\frac52})$. From the uniform estimates, we have $U\in L^\infty(0,\infty;B_{2,1}^{\frac52-2\alpha})\cap L^1(0,\infty;\dot B_{2,1}^{\frac52})$. Uniqueness can be proved similarly. We still have to prove that $U$ is continuous with values in $B_{2,1}^{\frac52-2\alpha}$ (so $u^2$ is continuous). $U^{n+1}$ satisfies

$$\partial_t U^{n+1}=-(U^{n+1}\cdot\nabla)u^1+(U^n\cdot\nabla)U^{n+1}-(u^1\cdot\nabla)U^{n+1}-\Lambda^{2\alpha}U^{n+1}-\nabla P^{n+1}.$$

Since the right-hand side of the above belongs to $L^1(0,\infty;B_{2,1}^{\frac52-2\alpha})$, we readily have that $U^{n+1}$ is continuous with values in $B_{2,1}^{\frac52-2\alpha}$. Since $U^n$ converges to $U$ in $L^\infty(0,\infty;B_{2,1}^{\frac52-2\alpha})$, $U$ is continuous with values in $B_{2,1}^{\frac52-2\alpha}$. Similarly to the proof of Theorem 1 and the a priori estimates of solution, we have $u^2\in C((0,\infty);B_{2,1}^{\frac52})$. This completes the proof of Theorem 2. □

*Proof of Theorem* 3. We follow closely [15].

Step 1. *A priori estimates*

If we set $u=e^{-t\Lambda^{2\alpha}}u_0+w$, then we have

$$\begin{cases}\partial_t w+\Lambda^{2\alpha}w=Q(e^{-t\Lambda^{2\alpha}}u_0,e^{-t\Lambda^{2\alpha}}u_0)+2Q(e^{-t\Lambda^{2\alpha}}u_0,w)+Q(w,w),\\ \qquad \mathbb{R}^3\times\mathbb{R}_+,\quad \frac12<\alpha<\frac54,\\ w(0,x)=0.\end{cases}$$

By using Proposition 2, we have for $\rho>\max\{\frac{2\alpha}{2\alpha-1},2\}$,

$$\begin{aligned}\|w\|_{L^\infty(\dot B_p)}&\le C\|Q(w,w)\|_{L^\infty(\dot B_p^{\frac3p+1-4\alpha})}\\&\quad+C\|Q(e^{-t\Lambda^{2\alpha}}u_0,w)\|_{\tilde L^\rho(\dot B_p^{\frac3p+1-2\alpha(\frac{2\rho-1}{\rho})})}\\&\quad+C\|Q(e^{-t\Lambda^{2\alpha}}u_0,e^{-t\Lambda^{2\alpha}}u_0)\|_{\tilde L^{\frac\rho2}(\dot B_p^{\frac3p+1-4\alpha\frac{\rho-1}{\rho}})}.\end{aligned}$$

By using Proposition 3, we have

$$\|Q(w,w)\|_{L^\infty(\dot B_p^{\frac3p+1-4\alpha})}\le C\|w\|^2_{L^\infty(\dot B_p)},$$

$$\|Q(e^{-t\Lambda^{2\alpha}}u_0,w)\|_{\tilde L^\rho(\dot B_p^{\frac3p+1-2\alpha(\frac{2\rho-1}{\rho})})}\le C\|e^{-t\Lambda^{2\alpha}}u_0\|_{\tilde L^\rho(\dot B_p^{\frac3p+1-2\alpha(\frac{\rho-1}{\rho})})}\|w\|_{L^\infty(\dot B_p)},\tag{28}$$

and

$$\|Q(e^{-t\Lambda^{2\alpha}}u_0,e^{-t\Lambda^{2\alpha}}u_0)\|_{\tilde L^{\frac\rho2}(\dot B_p^{\frac3p+1-4\alpha(\frac{\rho-1}{\rho})})}\le C\|e^{-t\Lambda^{2\alpha}}u_0\|^2_{\tilde L^\rho(\dot B_p^{\frac3p+1-2\alpha(\frac{\rho-1}{\rho})})}.\tag{29}$$

Using Proposition 2, we obtain

$$\|Q(e^{-t\Lambda^{2\alpha}}u_0, w)\|_{\tilde{L}^\rho(\dot{B}_p^{\frac{3}{p}+1-2\alpha(\frac{2\rho-1}{\rho})})} \leq C\|u_0\|_{\dot{B}_p}\|w\|_{L^\infty(\dot{B}_p)},$$

and

$$\|Q(e^{-t\Lambda^{2\alpha}}u_0, e^{-t\Lambda^{2\alpha}}u_0)\|_{\tilde{L}^{\frac{\rho}{2}}(\dot{B}_p^{\frac{3}{p}+1-4\alpha(\frac{\rho-1}{\rho})})} \leq C\|u_0\|^2_{\dot{B}_p}.$$

Thus we have

$$\begin{aligned}\|w\|_{L^\infty(\dot{B}_p)} &\leq C\|w\|^2_{L^\infty(\dot{B}_p)} + C\|u_0\|_{\dot{B}_p}\|w\|_{L^\infty(\dot{B}_p)} + C\|u_0\|^2_{\dot{B}_p} \\ &\leq C_{16}(\|w\|_{L^\infty(\dot{B}_p)} + \|u_0\|_{\dot{B}_p})^2. \end{aligned} \tag{30}$$

Step 2. *Iteration*

We define the iterating sequence

$$\begin{cases}\partial_t w_{n+1} + \Lambda^{2\alpha}w_{n+1} = Q(e^{-t\Lambda^{2\alpha}}u_0, e^{-t\Lambda^{2\alpha}}u_0) + 2Q(e^{-t\Lambda^{2\alpha}}u_0, w_n) + Q(w_n, w_n), \\ \qquad \mathbb{R}^3 \times \mathbb{R}_+, \frac{1}{2} < \alpha < \frac{5}{4}, \\ w_{n+1}(0,x) = 0. \end{cases}$$

Similar to apriori estimates, we have

$$\|w_{n+1}\|_{L^\infty(\dot{B}_p)} \leq C_{16}(\|w_n\|_{L^\infty(\dot{B}_p)} + \|u_0\|_{\dot{B}_p})^2.$$

Choose appropriately small $\epsilon$ and $w_1$ such that $\|u_0\|_{\dot{B}_p} < \epsilon$, $\|w_1\|_{L^\infty(\dot{B}_p)} < \epsilon$, and $4C_{16}\epsilon^2 \leq \epsilon$. Then we have $\|w_n\|_{L^\infty(\dot{B}_p)} \leq \epsilon$, for all $n$.

Step 3. *Equations of differences and existence*

Let $\delta w_{n+1} = w_{n+1} - w_n$. We have the following equation of the differences:

$$\begin{aligned}\partial_t \delta w_{n+1} + \Lambda^{2\alpha}\delta w_{n+1} &= 2Q(e^{-t\Lambda^{2\alpha}}u_0, \delta w_n) \\ &\quad + Q(w_n + w_{n-1}, \delta w_n).\end{aligned}$$

Similar to a priori estimates, we obtain

$$\begin{aligned}\|\delta w_{n+1}\|_{L^\infty(\dot{B}_p)} &\leq C\|u_0\|_{\dot{B}_p}\|\delta w_n\|_{L^\infty(\dot{B}_p)} \\ &\quad + C(\|w_n\|_{L^\infty(\dot{B}_p)} + \|w_{n-1}\|_{L^\infty(\dot{B}_p)})\|\delta w_n\|_{L^\infty(\dot{B}_p)}.\end{aligned}$$

Choosing $\epsilon$ so small that we obtain

$$\|\delta w_{n+1}\|_{L^\infty(\dot{B}_p)} \leq \frac{1}{2}\|\delta w_n\|_{L^\infty(\dot{B}_p)}.$$

By iteration, $w_n$ is a Cauchy sequence in $C([0,\infty);\dot{B}_p)$ and we have $w_n \to w$ in $C([0,\infty);\dot{B}_p)$. Since $u = e^{-t\Lambda^{2\alpha}}u_0 + w$, we obtain $u \in L^\infty(\dot{B}_p)$. We next prove uniqueness. If $e^{-t\Lambda^{2\alpha}}u_0 + w_1$ and $e^{-t\Lambda^{2\alpha}}u_0 + w_2$ are solutions of $(\text{SNS})_\alpha$ with $w_1, w_2 \in C([0,\infty);\dot{B}_p)$ and we set $\delta w = w_1 - w_2$, then we have the following estimates for any $T > 0$ similarly to the above:

$$\begin{aligned}\|\delta w\|_{L^\infty_T(\dot{B}_p)} &\leq C_{17}\|u_0\|_{\dot{B}_p}\|\delta w\|_{L^\infty_T(\dot{B}_p)} \\ &\quad + C_{18}(\|w_1\|_{L^\infty_T(\dot{B}_p)} + \|w_2\|_{L^\infty_T(\dot{B}_p)})\|\delta w\|_{L^\infty_T(\dot{B}_p)}.\end{aligned}$$

Choose $\epsilon$ so small that $C_{17}\epsilon \le \frac{1}{4}$ and $C_{18}\epsilon \le \frac{1}{4}$. Since $w_1$ and $w_2 \in C([0,\infty);\dot{B}_p)$, we can choose $T_1 > 0$ so small that

$$\|w_1\|_{L^\infty_{T_1}(\dot{B}_p)} + \|w_2\|_{L^\infty_{T_1}(\dot{B}_p)} \le \epsilon.$$

Since we obtain $\|\delta w\|_{L^\infty_{T_1}(\dot{B}_p)} \le \frac{1}{2}\|\delta w\|_{L^\infty_{T_1}(\dot{B}_p)}$, we immediately have $\delta w = 0$ on $[0, T_1]$. Doing the above process repeatedly on $[T_1, 2T_1]$, $[2T_1, 3T_1]$,..., we have $\delta w = 0$ on $[0,\infty)$. Thus we complete the proof of the unique existence. □

## Appendix

For the proof of Proposition 2, we follow the similar ideas in [15]. We first prove the following lemma, which is an extended version of Lemma 2.1 of [15].

**Lemma 1.** *There exists a constant $C > 0$ such that for any $t > 0$,*

$$\|\Delta_q e^{-t\Lambda^{2\alpha}} u\|_{L^p} \le \frac{1}{C} e^{-Ct2^{2q\alpha}} \|\Delta_q u\|_{L^p}. \tag{31}$$

*Proof.* Consider $\phi \in \mathcal{D}(\mathbb{R}^3 \setminus \{0\})$. Since we can assume that $u$ has support in an annulus, we obtain that

$$\begin{aligned} e^{-t\Lambda^{2\alpha}} u &= \phi(\lambda^{-1}D) e^{-t\Lambda^{2\alpha}} u \\ &= \mathcal{F}^{-1}(\phi(\lambda^{-1}\xi) e^{-t|\xi|^{2\alpha}} \hat{u}(\xi)) \\ &= \tilde{g}_\lambda(t,\cdot) * u, \end{aligned}$$

with $\tilde{g}_\lambda(t,x) = \int e^{i(x|\xi)} \phi(\lambda^{-1}\xi) e^{-t|\xi|^{2\alpha}} d\xi$. To prove the lemma, it is enough to show that

$$\|\tilde{g}_\lambda(t,\cdot)\|_{L^1} \le C e^{-Ct\lambda^{2\alpha}}.$$

Say, $\tilde{g}_\lambda(t,x) = \lambda^3 g_\lambda(t,\lambda x)$ with

$$g_\lambda(t,x) = \int e^{i(x|\xi)} \phi(\xi) e^{-t\lambda^{2\alpha}|\xi|^{2\alpha}} d\xi.$$

By calculation, we have

$$\begin{aligned} g_\lambda(t,x) &= (1+|x|^2)^{-3} \int (1+|x|^2)^3 e^{i(x|\xi)} \phi(\xi) e^{-t\lambda^{2\alpha}|\xi|^{2\alpha}} d\xi \\ &= (1+|x|^2)^{-3} \int (Id - \Delta_\xi)^3 (e^{i(x|\xi)}) \phi(\xi) e^{-t\lambda^{2\alpha}|\xi|^{2\alpha}} d\xi \\ &= (1+|x|^2)^{-3} \int e^{i(x|\xi)} (Id - \Delta)^3 (\phi(\xi) e^{-t\lambda^{2\alpha}|\xi|^{2\alpha}}) d\xi. \end{aligned}$$

The Leibnitz formula gives us that

$$(Id - \Delta)^3 (\phi(\xi) e^{-t\lambda^{2\alpha}|\xi|^{2\alpha}}) = \sum_{\substack{|\beta|\le 6 \\ \gamma \le \beta}} C_{\beta,\gamma} \partial^{(\beta-\gamma)} \phi(\xi) \partial^\gamma (e^{-t\lambda^{2\alpha}|\xi|^{2\alpha}}).$$

We have also the following formula by algebraic computation.

$$e^{t\lambda^{2\alpha}|\xi|^{2\alpha}}\partial^\gamma(e^{-t\lambda^{2\alpha}|\xi|^{2\alpha}}) = \sum_{\substack{\gamma_1+..+\gamma_m=\gamma\\ |\gamma_j|\geq 1}} (-t\lambda^{2\alpha})^m\Pi_{j=1}^m\partial^{\gamma_j}(|\xi|^{2\alpha}).$$

Since $\phi$ has support in an annulus, we have

$$\begin{aligned}|\partial^\gamma(e^{-t\lambda^{2\alpha}|\xi|^{2\alpha}})| &\leq C^{|\gamma|}(1+t\lambda^{2\alpha})^{|\gamma|}e^{-t\lambda^{2\alpha}|\xi|^{2\alpha}}\\ &\leq C^{|\gamma|}(1+t\lambda^{2\alpha})^{|\gamma|}e^{-ct\lambda^{2\alpha}}.\end{aligned}$$

Thus we have

$$|g_\lambda(t,x)| \leq (1+|x|^2)^{-3}e^{-Ct\lambda^{2\alpha}}.$$

The proof of lemma is completed. □

*Proof of Proposition* 2. First, we have the following estimate by Lemma 1.

$$\|e^{-t\Lambda^{2\alpha}}\Delta_q u_0\|_{L^p} \leq \frac{1}{C}e^{-Ct2^{2q\alpha}}\|\Delta_q u_0\|_{L^p}. \tag{32}$$

Taking $L^\rho$-norm over $[0,T]$ of both sides of (32), we have

$$\begin{aligned}\left(\int_0^T \|e^{-t\Lambda^{2\alpha}}\Delta_q u_0\|_{L^p}^\rho dt\right)^{\frac{1}{\rho}} &\leq C\left(\int_0^T e^{-Ct2^{2q\alpha}\rho}dt\right)^{\frac{1}{\rho}}\|\Delta_q u_0\|_{L^p}\\ &\leq C2^{-\frac{2q\alpha}{\rho}}\left(1-e^{-cT2^{2q\alpha}\rho}\right)^{\frac{1}{\rho}}\|\Delta_q u_0\|_{L^p}. \qquad (33)\end{aligned}$$

Multiplying with $2^{2q(s+\frac{2\alpha}{\rho})}$ and taking $l^r$ norm on both sides of (33), we obtain

$$\left(\sum_{q\in\mathbb{Z}} 2^{2q(s+\frac{2\alpha}{\rho})r}\|e^{-t\Lambda^{2\alpha}}\Delta_q u_0\|_{L_T^\rho(L^p)}^r\right)^{\frac{1}{r}} \leq C\left(\sum_{q\in\mathbb{Z}} 2^{2qsr}\|\Delta_q u_0\|_{L^p}^r\right)^{\frac{1}{r}}$$

Thus we proved the first part of Proposition 2.

Taking $\Delta_q$ operation on the equation in Proposition 2, we obtain

$$\partial_t\Delta_q u + \Lambda^{2\alpha}\Delta_q u = \Delta_q f.$$

Since $u_0 = 0$, we have

$$\Delta_q u(t) = \int_0^t e^{-(t-\tau)\Lambda^{2\alpha}}\Delta_q f(\tau)d\tau.$$

Taking $L^p$ norm, we have

$$\begin{aligned}\|\Delta_q u(t)\|_{L^p} &\leq C\int_0^t \|e^{-(t-\tau)\Lambda^{2\alpha}}\Delta_q f(\tau)\|_{L^p}d\tau\\ &\leq C\int_0^t e^{-C(t-\tau)2^{2q\alpha}}\|\Delta_q f(\tau)\|_{L^p}d\tau.\end{aligned}$$

In the above last inequality, we used the estimates of Lemma 1. Taking $L^{\rho_1}$ norm over $[0,T]$, we obtain the following by using Fubini's Theorem.

$$\begin{aligned}\left(\int_0^T \|\Delta_q u(t)\|_{L^p}^{\rho_1} dt\right)^{\frac{1}{\rho_1}} &\le C\left(\int_0^T \left(\int_0^t e^{-c(t-\tau)2^{2q\alpha}} \|\Delta_q f(\tau)\|_{L^p} d\tau\right)^{\rho_1} dt\right)^{\frac{1}{\rho_1}}\\ &\le C\int_0^T \left(\int_\tau^T e^{-c(t-\tau)2^{2q\alpha}\rho_1} dt\right)^{\frac{1}{\rho_1}} \|\Delta_q f(\tau)\|_{L^p} d\tau\\ &\le C\int_0^T 2^{-\frac{2q\alpha}{\rho_1}}\left(1-e^{-cT2^{2q\alpha}\rho_1}\right)^{\frac{1}{\rho_1}} e^{c\tau 2^{2q\alpha}} \|\Delta_q f(\tau)\|_{L^p} d\tau\\ &\le C2^{-\frac{2q\alpha}{\rho_1}} 2^{-2q\alpha(1-\frac{1}{\rho_2})}\left(\int_0^T \|\Delta_q f(\tau)\|_{L^p}^{\rho_2} d\tau\right)^{\frac{1}{\rho_2}}.\end{aligned}$$

In the above last inequality, we used Hölder's inequality and the following calculation:

$$\left(\int_0^T e^{c\tau 2^{2q\alpha}(\frac{\rho_2}{\rho_2-1})} d\tau\right)^{\frac{\rho_2-1}{\rho_2}} \le C2^{-2q\alpha(1-\frac{1}{\rho_2})}.$$

Multiplying with $2^{q(s+2\alpha(1+\frac{1}{\rho_1}-\frac{1}{\rho_2}))}$ on both sides of the inequality, we have

$$2^{q(s+2\alpha(1+\frac{1}{\rho_1}-\frac{1}{\rho_2}))}\|\Delta_q u\|_{L_T^{\rho_1}(L^p)} \le C2^{qs}\|\Delta_q f\|_{L_T^{\rho_2}(L^p)}.$$

Thus we have (3). This completes the proof of Proposition 2. □

**Acknowledgements**

This research is supported partially by the grant no.2002-2-10200-002-5 from the basic research program of the KOSEF, and J. Lee is supported by the BK 21 project.

## References

[1] H. Beirao da Veiga and P. Secchi, *$L^p$-stability for the strong solutions of the Navier–Stokes equations in the whole space,* Arch. Rat. Mech. Anal. **98** (1987), 65–70.

[2] G. Bourdaud, *Reálisations des espaces de Besov homogènes,* Arkiv för matematik **26** (1988), 41–54.

[3] M. Cannone, *Ondelettes, Paraproduits et Navier–Stokes,* Diderot Editeur, 1995.

[4] M. Cannone, *A generalization of a theorem by Kato on Navier–Stokes equations,* Rev. Mat. Iberoamericana **13** (1997), 515–541.

[5] M. Cannone and G. Karch, *Incompressible Navier–Stokes equations in abstract Banach spaces,* Tosio Kato's Method and Principle for Evolution equations in Mathematical Physics, (2001).

[6] M. Cannone and Y. Meyer, *Littlewood–Paley decomposition and Navier–Stokes equations,* Methods and Applications of Analysis **2** (1995), 307–319.

[7] M. Cannone, Y. Meyer, and F. Planchon, *Solutions auto-similaires des équations de Navier–Stokes in* $\mathbb{R}^3$, Exposé n. VIII, Séminaire X-EDP, Ecole Polytechnique (Janvier 1994).

[8] M. Cannone and F. Planchon, *On the nonstationary Navier–Stokes equations with an external force,* Advances in Differential Equations **4** (1999), 697–730.

[9] M. Cannone and F. Planchon, *On the regularity of the bilinear term for solutions to the incompressible Navier–Stokes equations* **16** (2000), 1–16.

[10] D. Chae, *On the Well-Posedness of the Euler Equations in the Triebel–Lizorkin Spaces,* Comm. Pure Appl. Math. **55** (2002), 654–678.

[11] D. Chae, *Local Existence and Blow-up Criterion for the Euler Equations in the Besov Spaces,* RIM-GARC Preprint no. 01-7.

[12] J.-Y. Chemin, *Remarques sur l'existence globale pour le système de Navier–Stokes incompressible*, SIAM J. Math. Anal. **23** (1992), 20–28.

[13] J.-Y. Chemin, *About Navier–Stokes system,* Prépublication du Laboratorie d'analyse numérique de Paris 6, R96023, 1996.

[14] J.-Y. Chemin, *Perfect incompressible fluids,* Clarendon Press, Oxford, (1998).

[15] J.-Y. Chemin, *Théorèmes d'unicité pour le système de Navier–Stokes tridimensionnel,* Journal d'Analyse Mathématique **77** (1999), 27–50.

[16] R. Danchin, *Global existence in critical spaces for compressible Navier–Stokes equations,* Invent. math. **141** (2000), 579–614.

[17] R. Danchin, *Local theory in critical spaces for compressible viscous and heat-conductive gases,* Comm. Partial Differential Equations **26** (2001), 1183–1233.

[18] M. Frazier, B. Jawerth and G. Weiss, *Littlewood–Paley theory and the study of function spaces,* AMS-CBMS Regional Conference Series in Mathematics **79** (1991).

[19] H. Fujita and T. Kato, *On the Navier–Stokes initial value problem I,* Arch. Rational Mech. Anal. **16** (1964), 269–315.

[20] T. Kato, *Strong* $L^p$ *solutions of the Navier–Stokes equations in* $\mathbb{R}^m$ *with applications to weak solutions,* Math. Zeit. **187** (1984), 471–480.

[21] H. Koch and D. Tataru, *Well-posedness for the Navier–Stokes equations,* Advances in Mathematics **157** (2001), 22–35.

[22] J.L. Lions, *Quelques méthodes de résolution des problèmes aux limites non linéaries,* Dunod, Paris, 1969.

[23] G. Ponce, R. Racke, T.C. Sideris, and E.S. Titi, *Global stability of large solutions to the 3D Navier–Stokes equations*, Commun. Math. Phys. **159** (1994), 329–341.

[24] H. Triebel, *Theory of Function Spaces,* Birkhäuser, 1983.

[25] M. Wiegner, *Decay and stability in* $L_p$ *for strong solutions of Cauchy problem for the Navier–Stokes equations,* The Navier–Stokes equations. Theory and numerical methods, Proceedings Oberwolfach (1988), eds. J.G. Heywood, et al., Lect. Notes Math. 1431, Berlin, Heidelberg, New York, Springer, 1990, 95–99.

[26] M. Vishik, *Hydrodynamics in Besov spaces,* Arch. Rational Mech. Anal **145** (1998), 197–214.

Dongho Chae† and Jihoon Lee††
School of Mathematical Sciences
Seoul National University
Seoul 151-747
Korea
e-mail: dhchae@math.snu.ac.kr†
zhlee@math.snu.ac.kr††

Advances in Mathematical Fluid Mechanics, 53–78


# The Commutation Error of the Space Averaged Navier–Stokes Equations on a Bounded Domain

A. Dunca[1], V. John[2] and W.J. Layton[3]

**Abstract.** In Large Eddy Simulation of turbulent flows, the Navier–Stokes equations are convolved with a filter and differentiation and convolution are interchanged, introducing an extra commutation error term, which is nearly universally dropped from the resulting equations. We show that the commutation error is asymptotically negligible in $L^p(\mathbb{R}^d)$ (i.e., it vanishes as the averaging radius $\delta \to 0$) if and only if the fluid and the boundary exert exactly zero force on each other. Next, we show that the commutation error tends to zero in $H^{-1}(\Omega)$ as $\delta \to 0$. Convergence is proven also for a weak form of the commutation error. The order of convergence is studied in both cases. Last, we study the influence of the commutation error on the energy balance of the filtered equations.

**Mathematics Subject Classification (2000).** 35Q30, 76F65.

**Keywords.** Large eddy simulation, commutation error.

## 1. Introduction

The space averaged Navier–Stokes equations for the space averaged fluid velocity $\overline{\mathbf{u}}$ and pressure $\overline{p}$ are the basic equations for large eddy simulation (LES) of turbulent flows. They are derived in many papers and in nearly every book on turbulence modeling, e.g. Aldama [2], Lesieur [20], Pope [21] or Sagaut [23], from the Navier–Stokes equations as follows :

1. One chooses a filter $g(\mathbf{x})$ and an averaging radius $\delta > 0$. The large eddies $\overline{\mathbf{u}}$ (of size $\geq O(\delta)$) are defined by filtering the underlying fluid velocity $\mathbf{u}$ :

$$\overline{\mathbf{u}} := g * \mathbf{u}.$$

[1]partially supported by NSF grants DMS 9972622, INT 9814115 and INT 9805563
[2]partially supported by the Deutsche Akademische Austauschdients (D.A.A.D.)
[3]partially supported by NSF grants DMS 9972622, INT 9814115 and INT 9805563

2. To derive the equations for $\overline{\mathbf{u}}$, the Navier–Stokes equations are convolved with $g(\cdot)$.
3. Ignoring boundaries and commuting convolution and differentiation leads to the space averaged Navier–Stokes equations, given by

$$\overline{\mathbf{u}}_t - \nabla \cdot \mathbb{S}(\overline{\mathbf{u}}, \overline{p}) + \nabla \cdot (\overline{\mathbf{u}\mathbf{u}^T}) = \overline{\mathbf{f}}, \quad \nabla \cdot \overline{\mathbf{u}} = 0, \tag{1}$$

where the stress tensor associated with the velocity and pressure averages $(\overline{\mathbf{u}}, \overline{p})$ is given by

$$\mathbb{S}(\overline{\mathbf{u}}, \overline{p}) := 2\nu\mathbb{D}(\overline{\mathbf{u}}) - \overline{p}\mathbb{I} \ \text{ where } \ \mathbb{D}(\overline{\mathbf{u}}) = \frac{\nabla\overline{\mathbf{u}} + \nabla\overline{\mathbf{u}}^T}{2} \tag{2}$$

is the velocity deformation tensor.

One central problem in LES is the closure problem of modeling $\nabla \cdot (\overline{\mathbf{u}\mathbf{u}^T})$ in terms of $\overline{\mathbf{u}}$, see, e.g., Sagaut [23]. We shall show herein that there is in fact another possibly serious closure problem in steps 2 and 3 above leading to the incorrect space filtered equations (1).

It is often reported in the LES literature that difficulties exist for simulating turbulence driven by interaction of flows with boundaries. In this report, we will show one reason: when the flow is given in a bounded domain with typical no–slip boundary conditions and the strong form of the space averaged Navier–Stokes equations is used, steps 2 and 3 lead to an $O(1)$ error near the boundary. A correct derivation of (1) (Section 2) reveals that an extra commutation error term $A_\delta(\mathbb{S}(\mathbf{u}, p))$, see Definition 2.1, must be included in (1). We show, Proposition 4.2, that $\|A_\delta(\mathbb{S}(\mathbf{u}, p))\|_{L^p(\mathbb{R}^d)} \to 0$ as $\delta \to 0$ if and only if the traction or Cauchy stress vector of the underlying flow is identically zero on the boundary of the domain ! In other words, the equations (1) are reasonable only for flows in which the domain's boundary exerts no influence on the flow.

An inspection of the proof of Proposition 4.3 reveals that the commutation error $A_\delta(\mathbb{S}(\mathbf{u}, p))$ is largest at the boundary and decays rapidly as one moves away from the boundary.

If the commutation error term is simply dropped and then the strong form of the space averaged Navier–Stokes equations is discretized, as by, e.g., a finite difference method, the results of Section 4 show that the error committed is $O(1)$. On the other hand, variational methods, such as finite element, spectral or spectral element methods, discretize the weak form of the relevant equations. These methods are known to depend on the size of the $H^{-1}$–norm of any omitted terms. We show in Section 5 that variational methods are possible since the $H^{-1}(\Omega)$–norm of the dropped commutation error does approach zero as $\delta \to 0$, Proposition 5.1.

Section 6 studies the weak form of the commutation error, $(A_\delta(\mathbb{S}(\mathbf{u}, p)), \mathbf{v})$ for $\mathbf{v}$ fixed. The third main result, Proposition 6.1, is that the weak form of the commutation error tends to zero as $\delta \to 0$. The order of convergence in two dimensions is at least $O(\delta^{1-\varepsilon})$ with arbitrary $\epsilon > 0$.

The issue of the commutation error has appeared occasionally in the engineering community, e.g. see Fureby and Tabor [9], Ghosal and Moin [12], or Vasilyev et al. [25]. Its critical importance is beginning to be realized, see Das and Moser [6]. One approach, [12, 25], has been to shrink the averaging radius $\delta(\mathbf{x})$ as $\mathbf{x}$ tends to the boundary of the domain; the correct boundary conditions are then clear : $\overline{\mathbf{u}} = \mathbf{0}$. This approach requires extra resolution and another commutation error due to the non–constant filter width occurs. This other commutation error is usually ignored in the engineering literature on the basis of a one–dimensional Taylor series estimation of it for very smooth functions. Interesting and important mathematical challenges remain for this approach as well.

Other special treatments of the near wall regions, such as near wall models, see [23, Section 9.2.2] for an overview, are common in LES to attempt to correct for the error. Recently, there are new approaches to LES without modeling, such as post processing [16] and the variational multiscale method by Hughes and co–workers [15].

## 2. The space averaged Navier-Stokes equations in a bounded domain

To derive the correct space averaged Navier–Stokes equations in a bounded domain, we will extend all functions to $\mathbb{R}^d$ and derive the equations satisfied by these extensions. Then, the new equations will be convolved.

We will always use standard notations for Sobolev and Lebesgue spaces, e.g. see Adams [1]. For vectors and tensors (matrices), we use standard matrix–vector notations.

Let $\Omega$ be a bounded domain in $\mathbb{R}^d$, $d = 2, 3$, with Lipschitz boundary $\partial\Omega$ with outward pointing unit normal $\mathbf{n}$ and $(d-1)$–dimensional measure $|\partial\Omega| < \infty$. We consider the incompressible Navier-Stokes equations with homogeneous Dirichlet boundary conditions

$$\begin{array}{rcll} \mathbf{u}_t - 2\nu\nabla\cdot\mathbb{D}(\mathbf{u}) + \nabla\cdot(\mathbf{u}\mathbf{u}^T) + \nabla p & = & \mathbf{f} & \text{in } (0,T)\times\Omega, \\ \nabla\cdot\mathbf{u} & = & 0 & \text{in } [0,T]\times\Omega, \\ \mathbf{u} & = & \mathbf{0} & \text{in } [0,T]\times\partial\Omega, \\ \mathbf{u}\,|_{t=0} & = & \mathbf{u}_0 & \text{in } \Omega, \\ \int_\Omega p\, d\mathbf{x} & = & 0 & \text{in } (0,T], \end{array} \tag{3}$$

where $\nu$ is the constant kinematic viscosity.

It will be helpful to recall that the stress tensor $\mathbb{S}(\mathbf{u}, p)$ is given by

$$\mathbb{S}(\mathbf{u}, p) := 2\nu\mathbb{D}(\mathbf{u}) - p\mathbb{I},$$

where $\mathbb{I}$ is the unit tensor, and that the normal stress / Cauchy stress / traction vector on $\partial\Omega$ is defined by $\mathbb{S}(\mathbf{u}, p)\mathbf{n}$.

Our analysis will require that solutions $(\mathbf{u}, p)$ of (3) are regular enough such that the normal stress has a well defined trace on the $\partial\Omega$ which belongs to some

Lebesgue space defined on $\partial\Omega$. We assume that

$$\begin{aligned} \mathbf{u} \in \left(H^2(\Omega) \cap H_0^1(\Omega)\right)^d, \quad & p \in H^1(\Omega) \cap \mathrm{L}_0^2(\Omega) \quad \text{for a.e. } t \in [0,T], \\ & \mathbf{u} \in \left(H^1((0,T))\right)^d \quad \text{for a.e. } \mathbf{x} \in \overline{\Omega}. \end{aligned} \tag{4}$$

**Lemma 2.1.** *If* (4) *holds then* $\mathbb{S}(\mathbf{u},p)\mathbf{n}$ *belongs to* $\left(H^{1/2}(\partial\Omega)\right)^d$. *In particular, for a.e.* $t \in (0,T]$, $\mathbb{S}(\mathbf{u},p)\mathbf{n} \in (L^q(\partial\Omega))^d$ *with* $1 \le q < \infty$ *if* $d = 2$ *and* $1 \le q \le 4$ *if* $d = 3$ *and*

$$\|\mathbb{S}(\mathbf{u},p)\mathbf{n}\|_{(L^q(\partial\Omega))^d} \le C\left(\nu\|\mathbf{u}\|_{(H^2(\Omega))^d} + \|p\|_{H^1(\Omega)}\right). \tag{5}$$

*Proof.* This follows from the usual trace theorem and embedding theorems, e.g., see Galdi [10, Chapter II, Theorem 3.1]. □

**Remark 2.1.** The result that $\mathbb{S}(\mathbf{u},p)\mathbf{n} \in (L^q(\partial\Omega))^d$ for $1 \le q < 4$ suffices for our purposes but it can be sharpened considerably. For example, Giga and Sohr [13, Theorem 3.1, p. 84] show that provided $\mathbf{f}$ is smooth enough and the initial condition $\mathbf{u}_0 \in \left(W^{2-2/s,s}(\Omega)\right)^d$, $s > 0$, holds, then for a.e. $t > 0$, $\mathbf{u}_t$ and $\nabla\cdot(\mathbf{u}\mathbf{u}^T)$ belong to $(L^q(\Omega))^d$ and further $\mathbb{S}(\mathbf{u},p)\mathbf{n} \in (L^q(\partial\Omega))$ for a.e. $t > 0$ when $3/q + 2/s = 4$.

In writing down an equation like (1), $\mathbf{f}$ must be extended off $\Omega$ and then $(\mathbf{u},p)$ must be extended compatible with the extension of $\mathbf{f}$. For $\overline{\mathbf{f}}$ to be computable, $\mathbf{f}$ is extended by zero off $\Omega$. Thus, $(\mathbf{u},p)$ must be extended by zero off $\Omega$, too. This extension is reasonable since $\mathbf{u} = \mathbf{0}$ on $\partial\Omega$. An extension of $\mathbf{u}$ off $\Omega$ as an $\left(H^2(\mathbb{R}^d)\right)^d$ function exists but is unknown, in particular since $\mathbf{u}$ is not known. Using this extension, instead of $\mathbf{u} \equiv \mathbf{0}$ on $\mathbb{R}^d \setminus \Omega$, would make the extension of $\mathbf{f}$ unknowable and hence $\overline{\mathbf{f}}$ uncomputable in (1). Thus, define

$$\mathbf{u} = \mathbf{0}, \quad \mathbf{u}_0 = \mathbf{0}, \quad p = 0 \quad \mathbf{f} = \mathbf{0} \quad \text{if } \mathbf{x} \notin \overline{\Omega}.$$

The extended functions possess the following regularities:

$$\begin{aligned} \mathbf{u} \in \left(H_0^1(\mathbb{R}^d)\right)^d, \quad & p \in \mathrm{L}_0^2(\mathbb{R}^d) \quad \text{for a.e. } t \in [0,T], \\ & \mathbf{u} \in \left(H^1((0,T))\right)^d \quad \text{for a.e. } \mathbf{x} \in \mathbb{R}^d. \end{aligned} \tag{6}$$

From (4) and (6) follow that the first order weak derivatives of the extended velocity $\mathbf{u}_t$, $\nabla\mathbf{u}$ ,$\nabla\cdot\mathbf{u}$ and $\nabla\cdot(\mathbf{u}\mathbf{u}^T)$ are well defined on $\mathbb{R}^d$, taking their indicated values in $\Omega$ and being identically zero off $\Omega$.

Since $\mathbf{u} \notin \left(H^2(\mathbb{R}^d)\right)^d$, $p \notin H^1(\mathbb{R}^d)$, the terms $\nabla\cdot\mathbb{D}(\mathbf{u})$ and $\nabla p$ must be defined in the sense of distributions. To this end, let $\varphi \in C_0^\infty(\mathbb{R}^d)$. Since $p \equiv 0$ on $\mathbb{R}^d \setminus \Omega$, we get

$$(\nabla p)(\varphi) := -\int_{\mathbb{R}^d} p(\mathbf{x})\nabla\varphi(\mathbf{x})\mathrm{d}\mathbf{x} = \int_\Omega \varphi(\mathbf{x})\nabla p(\mathbf{x})\mathrm{d}\mathbf{x} - \int_{\partial\Omega} \varphi(\mathbf{s})p(\mathbf{s})\mathbf{n}(\mathbf{s})\mathrm{d}\mathbf{s}. \tag{7}$$

In the same way, one obtains

$$\begin{aligned}\nabla \cdot \mathbb{D}(\mathbf{u})(\varphi) &:= -\int_{\mathbb{R}^d} \mathbb{D}(\mathbf{u})(\mathbf{x})\nabla\varphi(\mathbf{x})d\mathbf{x} \\ &= \int_{\Omega} \varphi(\mathbf{x})\nabla \cdot \mathbb{D}(\mathbf{u})(\mathbf{x})d\mathbf{x} - \int_{\partial\Omega} \varphi(\mathbf{s})\mathbb{D}(\mathbf{u})(\mathbf{s})\mathbf{n}(\mathbf{s})d\mathbf{s}.\end{aligned} \tag{8}$$

Both distributions have compact support. From (7) and (8) it follows that the extended functions $(\mathbf{u}, p)$ fulfill the following distributional form of the momentum equation:

$$\mathbf{u}_t - 2\nu\nabla\cdot\mathbb{D}(\mathbf{u}) + \nabla\cdot(\mathbf{u}\mathbf{u}^T) + \nabla p = \mathbf{f} + \int_{\partial\Omega} \Big(2\nu\mathbb{D}(\mathbf{u})(\mathbf{s})\mathbf{n}(\mathbf{s}) - p(\mathbf{s})\mathbf{n}(\mathbf{s})\Big)\varphi(\mathbf{s})d\mathbf{s}. \tag{9}$$

The correct space averaged Navier–Stokes equations are now derived by convolving (9) with a filter function $g(\mathbf{x}) \in \mathrm{C}^\infty(\mathbb{R}^d)$. Let $H(\varphi)$ be a distribution with compact support which has the form

$$H(\varphi) = -\int_{\mathbb{R}^d} f(\mathbf{x})\partial_\alpha\varphi(\mathbf{x})d\mathbf{x},$$

where $\partial_\alpha$ is the derivative of $\varphi$ with the multi-index $\alpha$. Then, $H * g \in C^\infty(\mathbb{R}^d)$, see Rudin [22, Theorem 6.35], where

$$\overline{H}(\mathbf{x}) = (H * g)(\mathbf{x}) := H(g(\mathbf{x} - \cdot)) = -\int_{\mathbb{R}^d} f(\mathbf{y})\partial_\alpha g(\mathbf{x} - \mathbf{y})d\mathbf{y}. \tag{10}$$

Applying the convolution with $g$ to (9), using the fact that convolution and differentiation commute on $\mathbb{R}^d$, Hörmander [14, Theorem 4.1.1], and convolving the extra term on the right hand side accordingly to (10), we obtain the space averaged momentum equation

$$\begin{aligned}&\overline{\mathbf{u}}_t - 2\nu\nabla \cdot \mathbb{D}(\overline{\mathbf{u}}) + \nabla \cdot (\overline{\mathbf{u}\mathbf{u}^T}) + \nabla\overline{p} \\ &= \overline{\mathbf{f}} + \int_{\partial\Omega} g(\mathbf{x} - \mathbf{s})\left[2\nu\mathbb{D}(\mathbf{u})(\mathbf{s})\mathbf{n}(\mathbf{s}) - p(\mathbf{s})\mathbf{n}(\mathbf{s})\right] ds \quad \text{in } (0, T] \times \mathbb{R}^d.\end{aligned} \tag{11}$$

**Remark 2.2.** If the viscous term in the Navier–Stokes equations is written as $\nu\Delta\mathbf{u}$ instead of $2\nu\nabla \cdot \mathbb{D}(\mathbf{u})$, the resulting space averaged equation is given by replacing $2\nu\mathbb{D}(\overline{\mathbf{u}})$ in (11) by $\nu\nabla\overline{\mathbf{u}}$.

**Definition 2.1.** *The commutation error $A_\delta(\mathbb{S}(\mathbf{u}, p))$ in the space averaged Navier–Stokes equations is defined to be*

$$A_\delta(\mathbb{S}(\mathbf{u}, p)) := \int_{\partial\Omega} g(\mathbf{x} - \mathbf{s})(\mathbb{S}(\mathbf{u}, p)\mathbf{n})(\mathbf{s})ds.$$

The correct space averaged Navier–Stokes equations arising from the Navier–Stokes equations on a bounded domain thus possess an extra boundary integral, $A_\delta(\mathbb{S}(\mathbf{u}, p))$. Omitting this integral results in a commutation error. Including this integral in (1) introduces a new modeling question since it depends on the unknown normal stress on $\partial\Omega$ of $(\mathbf{u}, p)$ and not of $(\overline{\mathbf{u}}, \overline{p})$.

## 3. The Gaussian filter

We will present the results in the following sections for the Gaussian filter. This filter fits into the framework of Section 2. We shall briefly present the filter's properties that are used in the subsequent analysis in this section.

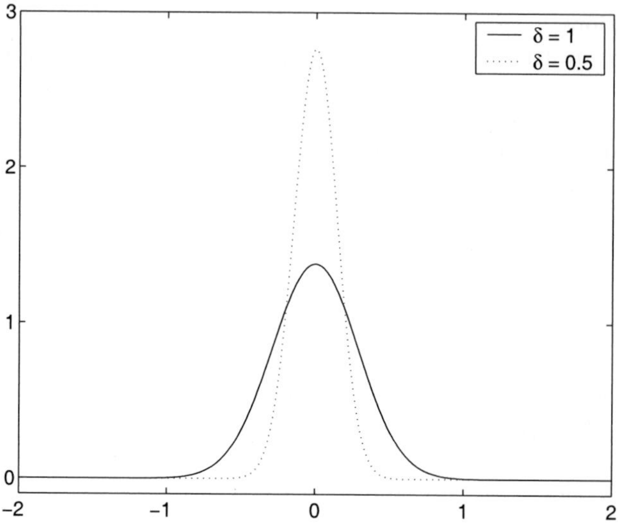


FIGURE 1. The Gaussian filter function in one dimension for different $\delta$

The Gaussian filter function has the form

$$g_\delta(\mathbf{x}) = \left(\frac{6}{\delta^2\pi}\right)^{d/2} \exp\left(-\frac{6}{\delta^2}\|\mathbf{x}\|_2^2\right),$$

see Figure 1, where $\|\cdot\|_2$ denotes the Euclidean norm of $\mathbf{x} \in \mathbb{R}^d$ and $\delta$ is a user–chosen positive length scale. The Gaussian filter has the following properties, which are easy to verify:

- regularity: $g_\delta \in C^\infty(\mathbb{R}^d)$,
- positivity: $0 < g_\delta(\mathbf{x}) \le \left(\frac{6}{\delta^2\pi}\right)^{\frac{d}{2}}$,
- integrability: $\|g_\delta\|_{L^p(\mathbb{R}^d)} < \infty$, $1 \le p \le \infty$, $\|g_\delta\|_{L^1(\mathbb{R}^d)} = 1$,
- symmetry: $g_\delta(\mathbf{x}) = g_\delta(-\mathbf{x})$,
- monotonicity: $g_\delta(\mathbf{x}) \ge g_\delta(\mathbf{y})$ if $\|\mathbf{x}\|_2 \le \|\mathbf{y}\|_2$.

**Lemma 3.1.**

i) *Let $\varphi \in L^p(\mathbb{R}^d)$, then for $1 \le p < \infty$,*

$$\lim_{\delta\to 0} \|g_\delta * \varphi - \varphi\|_{L^p(\mathbb{R}^d)} = 0.$$

ii) *Let* $\varphi \in L^\infty(\mathbb{R}^d)$. *If* $\varphi$ *is uniformly continuous on a set* $\omega$, *then* $g_\delta * \varphi \to \varphi$ *uniformly on* $\omega$ *as* $\delta \to 0$.

iii) *If* $\varphi \in C_0^\infty(\mathbb{R}^d)$, *then for* $1 \le p < \infty$, $0 \le r < \infty$,

$$\lim_{\delta \to 0} \|g_\delta * \varphi - \varphi\|_{W^{r,p}(\mathbb{R}^d)} = 0.$$

*Proof.* The proof of the first two statements can be found, e.g. in Folland [8, Theorem 0.13]. The third statement is an immediate consequence of the first one. □

For convenience, the Gaussian filter function with a scalar argument $x$ is understood in the following to be

$$g_\delta(x) := \left(\frac{6}{\delta^2 \pi}\right)^{\frac{d}{2}} \exp\left(-\frac{6x^2}{\delta^2}\right).$$

## 4. Error estimates in the $(L^p(\mathbb{R}^d))^d$–norm of the commutation error term

In this section, it is shown that the commutation error $A_\delta(\mathbb{S}(\mathbf{u}, p))$ belongs to $(L^p(\mathbb{R}^d))^d$. We show that $A_\delta(\mathbb{S}(\mathbf{u}, p))$ vanishes as $\delta \to 0$ if and only if the normal stress is identically zero a.e. on $\partial\Omega$. As noted earlier, this condition means the wall have zero influence on the wall-bounded turbulent flow. Thus, it is not expected to be satisfied in any interesting flow problem !

In view of Definition 2.1 and Lemma 2.1, it is necessary to study terms of the form

$$\int_{\partial\Omega} g_\delta(\mathbf{x} - \mathbf{s})\psi(\mathbf{s})d\mathbf{s} \tag{12}$$

with $\psi \in L^q(\partial\Omega)$, $1 \le q \le \infty$. We will first show, that (12) belongs to $L^p(\mathbb{R}^d), 1 \le p \le \infty$.

**Proposition 4.1.** *Let* $\psi \in L^q(\partial\Omega)$, $1 \le q \le \infty$, *then* (12) *belongs to* $L^p(\mathbb{R}^d), 1 \le p \le \infty$.

*Proof.* By the Cauchy–Schwarz inequality, one obtains with $r^{-1} + q^{-1} = 1$, $q > 1$,

$$\begin{aligned}\left|\int_{\partial\Omega} g_\delta(\mathbf{x} - \mathbf{s})\psi(\mathbf{s})d\mathbf{s}\right| &\le \left(\int_{\partial\Omega} g_\delta^r(\mathbf{x} - \mathbf{s})d\mathbf{s}\right)^{1/r} \|\psi\|_{L^q(\partial\Omega)} \\ &= \left(\int_{\partial\Omega} \left(\frac{6}{\delta^2\pi}\right)^{rd/2} \exp\left(-\frac{6r}{\delta^2}\|\mathbf{x} - \mathbf{s}\|_2^2\right) d\mathbf{s}\right)^{1/r} \|\psi\|_{L^q(\partial\Omega)}.\end{aligned}$$

As $2\|\mathbf{x} - \mathbf{s}\|_2^2 \ge \|\mathbf{x}\|_2^2 - 2\|\mathbf{s}\|_2^2$, it follows that

$$\exp\left(-\frac{6r\|\mathbf{x} - \mathbf{s}\|_2^2}{\delta^2}\right) \le \exp\left(3r\frac{-\|\mathbf{x}\|_2^2 + 2\|\mathbf{s}\|_2^2}{\delta^2}\right),$$

and

$$\begin{aligned}
&\left|\int_{\partial\Omega} g_\delta(\mathbf{x}-\mathbf{s})\psi(\mathbf{s})d\mathbf{s}\right| \\
&\quad \le \left(\frac{6}{\delta^2\pi}\right)^{d/2} \|\psi\|_{L^q(\partial\Omega)} \left(\int_{\partial\Omega} \exp\left(\frac{6r\|\mathbf{s}\|_2^2}{\delta^2}\right) d\mathbf{s}\right)^{1/r} \exp\left(-\frac{3\|\mathbf{x}\|_2^2}{\delta^2}\right) \qquad (13)\\
&\quad < \infty,
\end{aligned}$$

since $\partial\Omega$ is compact and the exponential is a bounded function. This proves the statement for $L^\infty(\mathbb{R}^d)$. The proof for $p \in [1,\infty)$ is obtained by raising both sides of (13) to the power $p$, integrating on $\mathbb{R}^d$ and using

$$\int_{\mathbb{R}^d} \exp\left(-\frac{3p\|\mathbf{x}\|_2^2}{\delta^2}\right) d\mathbf{x} < \infty.$$

If $q = 1$, we have for $1 \le p < \infty$

$$\begin{aligned}
\int_{\mathbb{R}^d} \left|\int_{\partial\Omega} g_\delta(\mathbf{x}-\mathbf{s})\psi(\mathbf{s})d\mathbf{s}\right|^p d\mathbf{x} &\le \int_{\mathbb{R}^d} \sup_{\mathbf{s}\in\partial\Omega} g_\delta^p(\mathbf{x}-\mathbf{s})d\mathbf{x}\, \|\psi\|_{L^1(\partial\Omega)}^p \\
&= \int_{\mathbb{R}^d} g_\delta^p(d(\mathbf{x},\partial\Omega))d\mathbf{x}\, \|\psi\|_{L^1(\partial\Omega)}^p.
\end{aligned}$$

We choose a ball $B(\mathbf{0},R)$ with radius $R$ such that $d(\mathbf{x},\partial\Omega) > \|\mathbf{x}\|_2/2$ for all $\mathbf{x} \notin B(\mathbf{0},R)$. Then, the integral on $\mathbb{R}^d$ is split into a sum of two integrals. The first integral is computed on $B(\mathbf{0},R)$. This is finite since the integrand is a continuous function on $\overline{B}(\mathbf{0},R)$. The second integral on $\mathbb{R}^d \setminus B(\mathbf{0},R)$ is also finite because

$$\int_{\mathbb{R}^d\setminus B(\mathbf{0},R)} g_\delta^p(d(\mathbf{x},\partial\Omega))d\mathbf{x} \le \int_{\mathbb{R}^d} g_\delta^p\left(\frac{\|\mathbf{x}\|_2}{2}\right) d\mathbf{x}$$

and the integrability of the Gaussian filter. This concludes the proof for $p < \infty$. For $p = \infty$, we have

$$\begin{aligned}
&\operatorname*{ess\,sup}_{\mathbf{x}\in\mathbb{R}^d} \left|\int_{\partial\Omega} g_\delta(\mathbf{x}-\mathbf{s})\psi(\mathbf{s})d\mathbf{s}\right| \\
&\quad \le \operatorname*{ess\,sup}_{\mathbf{x}\in\mathbb{R}^d} \operatorname*{ess\,sup}_{\mathbf{s}\in\partial\Omega} g_\delta(\mathbf{x}-\mathbf{s}) \|\psi\|_{L^1(\partial\Omega)} \le g_\delta(\mathbf{0})\|\psi\|_{L^1(\partial\Omega)} < \infty.
\end{aligned}$$

□

In the next proposition, we study the behaviour of the $L^p(\mathbb{R}^d)$–norm of (12) for $\delta \to 0$.

**Proposition 4.2.** *Let $\psi \in L^p(\partial\Omega)$, $1 \le p \le \infty$. A necessary and sufficient condition for*

$$\lim_{\delta\to 0} \left\|\int_{\partial\Omega} g_\delta(\mathbf{x}-\mathbf{s})\psi(\mathbf{s})d\mathbf{s}\right\|_{L^p(\mathbb{R}^d)} = 0, \qquad (14)$$

*$1 \le p \le \infty$, is that $\psi$ vanishes almost everywhere on $\partial\Omega$.*

*Proof.* It is obvious that the condition is sufficient.

Let (14) hold. From Hölder's inequality, we obtain for an arbitrary function $\varphi \in C_0^\infty(\mathbb{R}^d)$,

$$\begin{aligned}
&\lim_{\delta\to 0}\left|\int_{\mathbb{R}^d}\varphi(\mathbf{x})\left(\int_{\partial\Omega} g_\delta(\mathbf{x}-\mathbf{s})\psi(\mathbf{s})d\mathbf{s}\right)d\mathbf{x}\right| \\
&\quad\le \lim_{\delta\to 0}\|\varphi\|_{L^q(\mathbb{R}^d)}\left\|\int_{\partial\Omega} g_\delta(\mathbf{x}-\mathbf{s})\psi(\mathbf{s})d\mathbf{s}\right\|_{L^p(\mathbb{R}^d)} = 0 \qquad (15)
\end{aligned}$$

where $p^{-1}+q^{-1}=1$. By Fubini's theorem and the symmetry of the Gaussian filter, we have

$$\begin{aligned}
&\lim_{\delta\to 0}\int_{\mathbb{R}^d}\varphi(\mathbf{x})\left(\int_{\partial\Omega} g_\delta(\mathbf{x}-\mathbf{s})\psi(\mathbf{s})d\mathbf{s}\right)d\mathbf{x} \\
&\quad= \lim_{\delta\to 0}\int_{\partial\Omega}\psi(\mathbf{s})\left(\int_{\mathbb{R}^d} g_\delta(\mathbf{x}-\mathbf{s})\varphi(\mathbf{x})d\mathbf{x}\right)d\mathbf{s} = \int_{\partial\Omega}\psi(\mathbf{s})\varphi(\mathbf{s})d\mathbf{s}.
\end{aligned}$$

The last step is a consequence of Lemma 3.1 since $\varphi \in L^\infty(\mathbb{R}^d)$ and $\varphi$ is uniformly continuous on the compact set $\partial\Omega$. Thus, from (15) follows

$$0 = \left|\int_{\partial\Omega}\psi(\mathbf{s})\varphi(\mathbf{s})d\mathbf{s}\right|$$

for every $\varphi \in C_0^\infty(\mathbb{R}^d)$. This is true if and only if $\psi(\mathbf{s})$ vanishes almost everywhere on $\partial\Omega$. □

We will now bound the $L^p(\mathbb{R}^d)$-norm of (12) in terms of $\delta$. The next lemma proves a geometric property which is needed later.

**Lemma 4.1.** *Let $\Omega \subset \mathbb{R}^d$, $d = 2,3$ be a bounded domain with Lipschitz boundary $\partial\Omega$. Then there exists a constant $C > 0$ such that*

$$\left|\{\mathbf{x}\in\mathbb{R}^d | d(\mathbf{x},\partial\Omega)\le y\}\right| \le C(y+y^d) \qquad (16)$$

*for every $y \ge 0$, where $|\cdot|$ denotes the measure in $\mathbb{R}^d$.*

*Proof.* For simplicity, we present the proof for $\Omega$ being a simply connected domain. The analysis can be extended to the case that $\partial\Omega$ consists of a finite number of non-connected parts.

We will start with the case $d = 2$. We fix a point $\mathbf{x}_0$ on $\partial\Omega$ and an orientation of the boundary. Next, we construct $\mathbf{x}_1$ such that the length of the curve between $\mathbf{x}_0$ and $\mathbf{x}_1$ is $y$. Continuing this construction, we obtain a sequence $(\mathbf{x}_i)_{0\le i\le N}$ such that for every $0 \le i < N$ the length of curve between $\mathbf{x}_i$ and $\mathbf{x}_{i+1}$ is $y$. The length of the curve between $\mathbf{x}_N$ and $\mathbf{x}_0$ is less or equal than $y$, see Figure 2. The number of intervals is $N+1$ with $N < |\partial\Omega|/y \le N+1$. Obviously, we have

$$\{\mathbf{x}\in\mathbb{R}^d | d(\mathbf{x},\partial\Omega)\le y\} = \bigcup_{\mathbf{x}\in\partial\Omega}\overline{B}(\mathbf{x},y).$$

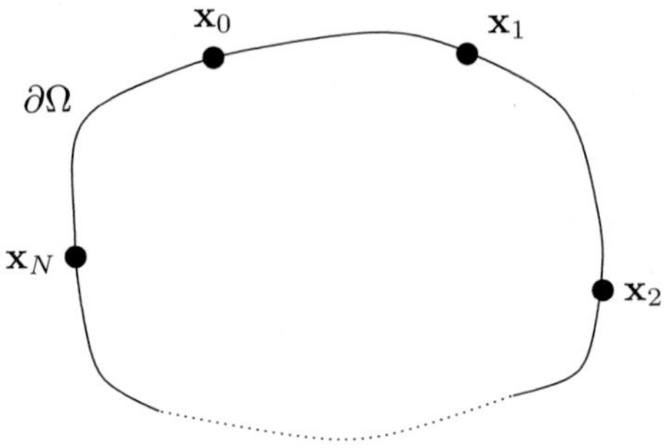


FIGURE 2. Mesh on $\partial\Omega$ for $d = 2$

But for every $\mathbf{x}$ in $\partial\Omega$, there exists an $i$ such that $\mathbf{x}$ is on the part of the curve from $\mathbf{x}_i$ to $\mathbf{x}_{i+1}$ or from $\mathbf{x}_N$ to $\mathbf{x}_0$. By the triangle inequality, this implies $\overline{B}(\mathbf{x}, y) \subset \overline{B}(\mathbf{x}_i, 2y)$. Thus

$$\{\mathbf{x} \in \mathbb{R}^d | d(x, \partial\Omega) \leq y\} \subset \bigcup_{0\leq i\leq N} \overline{B}(\mathbf{x}_i, 2y),$$

from which

$$\begin{aligned} |\{\mathbf{x} \in \mathbb{R}^d | d(\mathbf{x}, \partial\Omega) \leq y\}| &\leq \sum_{i=0}^{N} |\overline{B}(\mathbf{x}_i, 2y)| < \left(\frac{|\partial\Omega|}{y} + 1\right) 4\pi y^2 \\ &= 4\pi|\partial\Omega|y + 4\pi y^2 \end{aligned}$$

follows.

In the case $d = 3$, $\partial\Omega$ is a compact manifold. Then, for every $\mathbf{x} \in \partial\Omega$, there exists a neighborhood $U_\mathbf{x} \subset \partial\Omega$ such that its closure $\overline{U}_\mathbf{x}$ is homeomorphic to a closed square $\overline{V}_\mathbf{x} \subset \mathbb{R}^2$ through the homeomorphism $\phi_\mathbf{x} : \overline{V}_\mathbf{x} \to \overline{U}_\mathbf{x}$. The homeomorphism is Lipschitz continuous with a constant $L$. We cover the manifold by

$$\partial\Omega = \bigcup_{\mathbf{x}\in\partial\Omega} U_\mathbf{x}$$

and, because $\partial\Omega$ is compact, we can choose a finite cover $(U_{\mathbf{x}_i})_{0\leq i\leq N}$ which will be fixed. Let the length of the sides of $\overline{V}_{\mathbf{x}_i}$ be equal to $a_i$. We create a mesh over on $\overline{V}_{\mathbf{x}_i}$ of cells of size $y/L$ (or smaller). On this mesh, there are less than $(a_iL/y+2)^2$ vertices and we denote them by $(\mathbf{z}_j)_{0\leq j\leq P_i}$ where $P_i < (a_iL/y+2)^2$. The order of the vertices is not important. Next, we consider $\mathbf{z} \in U_{\mathbf{x}_i}$. Then, we find the closest vertex on the mesh to $\phi^{-1}(\mathbf{z})$ and denote it by $\mathbf{z}_k$. It is easy to see that

$$||\mathbf{z}_k - \phi^{-1}(\mathbf{z})||_2 \leq \frac{y}{L}$$

and the Lipschitz continuity of $\phi$ gives

$$||\phi(\mathbf{z}_k) - \mathbf{z}||_2 \leq L||\mathbf{z}_k - \phi^{-1}(\mathbf{z})||_2 \leq y.$$

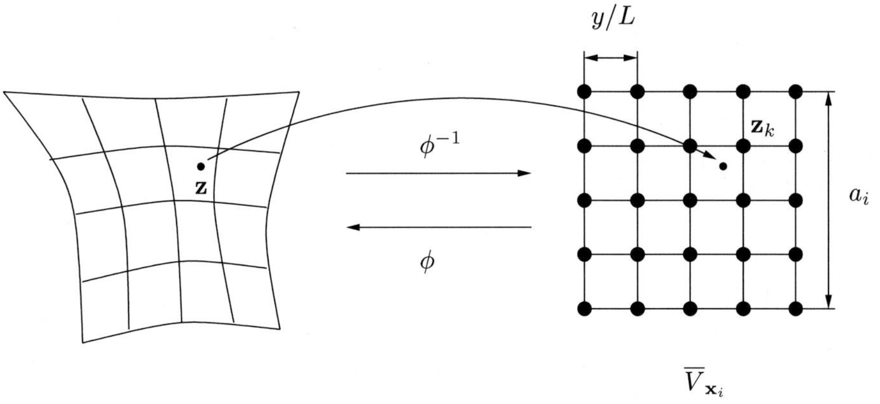


FIGURE 3. Homeomorphic map to the square $\overline{V}_{\mathbf{x}_i}$, $d = 3$

By the triangle inequality follows now

$$B(\mathbf{z}, y) \subset B(\phi(\mathbf{z}_k), 2y). \tag{17}$$

Because $\mathbf{z} \in U_{\mathbf{x}_i}$ was chosen arbitrary, for every $\mathbf{z} \in U_{\mathbf{x}_i}$ there exists $\mathbf{z}_k \in \overline{V}_{\mathbf{x}_i}$ such that (17) holds. Combining (17) for $U_{\mathbf{x}_i}$, $0 \le i \le N$, gives

$$\{\mathbf{x} \in \mathbb{R}^3 | d(\mathbf{x}, \partial\Omega) \le y\} \subset \bigcup_{0 \le i \le N} \bigcup_{0 \le k \le P_i} \overline{B}(\phi(\mathbf{z}_k), 2y).$$

By the sub–additivity and monotonicity of Lebesgue measure, we obtain

$$\begin{aligned} |\{\mathbf{x} \in \mathbb{R}^3 | d(\mathbf{x}, \partial\Omega) \le y\}| &\le \sum_{i=0}^{N} \sum_{k=0}^{P_i} |\overline{B}(\phi(z_k), 2y)| \le \sum_{i=0}^{N} \left( \frac{a_i L}{y} + 2 \right)^2 \frac{4}{3} \pi y^3 \\ &\le C(y^3 + y) \end{aligned}$$

for an appropriately chosen positive constant $C$. Note, the quadratic term in $y$ can be absorbed into the linear term for $y \le 1$ and into the cubic term for $y > 1$. □

**Proposition 4.3.** *Let $\Omega$ be a bounded domain in $\mathbb{R}^d$ with Lipschitz boundary $\partial\Omega$, $\psi \in L^p(\partial\Omega)$ for some $p > 1$ and $p^{-1} + q^{-1} = 1$. Then for every $\alpha \in (0, 1)$ and $k \in (0, \infty)$ there exist constants $C > 0$ and $\epsilon > 0$ such that*

$$\int_{\mathbb{R}^d} \left| \int_{\partial\Omega} g_\delta(\mathbf{x} - \mathbf{s}) \psi(\mathbf{s}) d\mathbf{s} \right|^k d\mathbf{x} \le C \delta^{1 + k\left(\frac{(d-1)\alpha}{q} - d\right)} \|\psi\|^k_{L^p(\partial\Omega)} \tag{18}$$

*for every $\delta \in (0, \epsilon)$ where $C$ and $\epsilon$ depend on $\alpha, k$ and $|\partial\Omega|$.*

*Proof.* We fix an $\alpha \in (0, 1)$. From Hölder's inequality, we obtain

$$\int_{\mathbb{R}^d} \left| \int_{\partial\Omega} g_\delta(\mathbf{x} - \mathbf{s}) \psi(\mathbf{s}) d\mathbf{s} \right|^k d\mathbf{x} \le \int_{\mathbb{R}^d} \left( \int_{\partial\Omega} g^q_\delta(\mathbf{x} - \mathbf{s}) d\mathbf{s} \right)^{k/q} d\mathbf{x} \; \|\psi\|^k_{L^p(\partial\Omega)}.$$

Let $B(\mathbf{x},\delta^\alpha)$ be the ball centered at $\mathbf{x}\in\mathbb{R}^d$ and with radius $\delta^\alpha$. Then, the term containing the Gaussian filter function can be estimated by the triangle inequality

$$\int_{\mathbb{R}^d}\left(\int_{\partial\Omega} g_\delta^q(\mathbf{x}-\mathbf{s})\mathrm{d}\mathbf{s}\right)^{k/q}\mathrm{d}\mathbf{x} \le C(k)\left(\int_{\mathbb{R}^d} B_\delta^k(\mathbf{x})\mathrm{d}\mathbf{x} + \int_{\mathbb{R}^d} C_\delta^k(\mathbf{x})\mathrm{d}\mathbf{x}\right) \tag{19}$$

where

$$B_\delta(\mathbf{x}) = \left(\int_{\partial\Omega\cap B(\mathbf{x},\delta^\alpha)} g_\delta^q(\mathbf{x}-\mathbf{s})\mathrm{d}\mathbf{s}\right)^{1/q}, C_\delta(\mathbf{x}) = \left(\int_{\partial\Omega\setminus B(\mathbf{x},\delta^\alpha)} g_\delta^q(\mathbf{x}-\mathbf{s})\mathrm{d}\mathbf{s}\right)^{1/q}$$

with the constant $C(k)$ depending only on $k$. We estimate the terms in (19) separately.

Using the monotonicity of the Gaussian filter, one can obtain the following inequality

$$C_\delta^k(\mathbf{x}) \le C\begin{cases} g_\delta^k(\delta^\alpha) & \text{if } d(\mathbf{x},\partial\Omega) < \delta^\alpha, \\ g_\delta^k(d(\mathbf{x},\partial\Omega)) & \text{if } d(\mathbf{x},\partial\Omega) \ge \delta^\alpha, \end{cases}$$

where $C = C(|\partial\Omega|)$. We refer to the function behind the brace as bounding function, see Figure 4 for a sketch in a special situation.

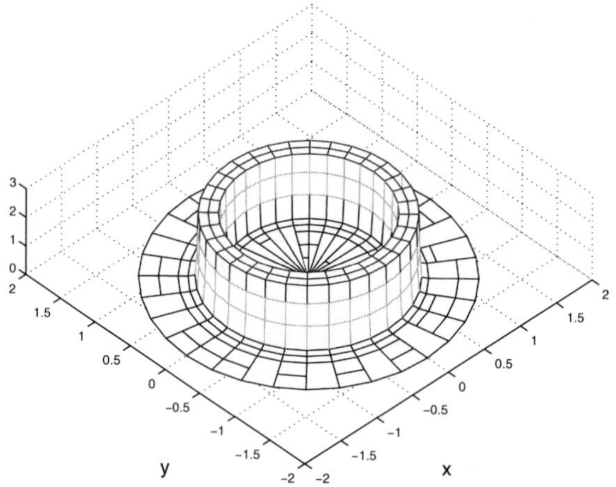


FIGURE 4. Bounding function of $C_\delta^k(\mathbf{x})$, $d=2$, $\partial\Omega = B(\mathbf{0},1)$, $\delta = 0.1$, $\alpha = 0.99$, $k=1$, $C = 2\pi$

Let $C(t) = \{(\mathbf{z},t) | d(\mathbf{z},\partial\Omega) \le y, t = g_\delta^k(y), \delta^\alpha \le y < \infty\}$ be the cross section of the bounding function at the function value $t$ and $A(t) = |C(t)|$ the area of the cross section. Then

$$\int_{\mathbb{R}^d} C_\delta^k(\mathbf{x})\mathrm{d}\mathbf{x} \le C\int_0^{g_\delta^k(\delta^\alpha)} A(t)\mathrm{d}t.$$

From Lemma 4.1, we know $A(t) \leq C(y^d + y)$, with $C$ depending only on $\Omega$. Using $g_\delta^k(y) = t$, changing variables and integrating by parts yield

$$\begin{aligned}\int_0^{g_\delta^k(\delta^\alpha)} A(t)\mathrm{d}t &\leq C\int_0^{g_\delta^k(\delta^\alpha)} (y^d + y)\mathrm{d}t = C\int_\infty^{\delta^\alpha} (y^d + y)\frac{\mathrm{d}}{\mathrm{d}y}(g_\delta^k(y))\mathrm{d}y \\ &= C\left((\delta^{d\alpha} + \delta^\alpha)g_\delta^k(\delta^\alpha) - d\int_\infty^{\delta^\alpha} y^{d-1}g_\delta^k(y)\mathrm{d}y - \int_\infty^{\delta^\alpha} g_\delta^k(y)\mathrm{d}y\right).\end{aligned}$$

The integrals on the last line will be estimated using the change of variables $y = \delta/t$ and by monotonicity considerations of the arising integrand. For $\delta$ sufficiently small, one obtains

$$\int_{\mathbb{R}^d} C_\delta^k(\mathbf{x})\mathrm{d}\mathbf{x} \leq C\left(\delta^{d(\alpha-k)} + \delta^{\alpha-kd}\right)\exp\left(-\frac{6k}{\delta^{2(1-\alpha)}}\right),$$

from what follows, since $\alpha < 1$,

$$\lim_{\delta\to 0}\int_{\mathbb{R}^d} C_\delta^k(\mathbf{x})\mathrm{d}\mathbf{x} = 0.$$

Now we will bound the second term in (19). The function $B_\delta^k(\mathbf{x})$ can be estimated from above in the following way

$$B_\delta^k(\mathbf{x}) \leq \begin{cases} |\partial\Omega \cap B(\mathbf{x},\delta^\alpha)|^{\frac{k}{q}} g_\delta^k(d(\mathbf{x},\partial\Omega)) & \text{if } d(\mathbf{x},\partial\Omega) < \delta^\alpha, \\ 0 & \text{if } d(\mathbf{x},\partial\Omega) \geq \delta^\alpha, \end{cases}$$

see Figure 5 for an illustration of the bounding function in a special situation. The bounding function is discontinuous, having a jump from the value 0 to the value $Cg_\delta^k(\delta^\alpha)$ at $\{\mathbf{x} \in \mathbb{R}^d \mid d(\mathbf{x},\partial\Omega) = \delta^\alpha\}$.

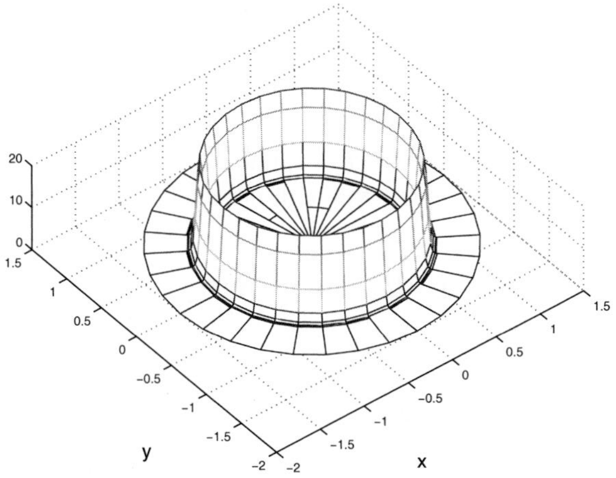


FIGURE 5. Bounding function of $B_\delta^k(\mathbf{x})$, $d = 2$, $\partial\Omega = B(\mathbf{0}, 1)$, $\delta = 0.1$, $\alpha = 0.99$, $k = 1$, $C = \delta^\alpha$

Since $\partial\Omega$ is smooth, we have $|\partial\Omega \cap B(\mathbf{x}, \delta^\alpha)| \le C\delta^{(d-1)\alpha}$ if $\delta$ is small enough. It follows

$$\int_{\mathbb{R}^d} B_\delta^k(\mathbf{x})\mathrm{d}\mathbf{x} \le C \int_{\{d(\mathbf{x},\partial\Omega)<\delta^\alpha\}} \delta^{\frac{(d-1)\alpha k}{q}} g_\delta^k(d(\mathbf{x},\partial\Omega))\mathrm{d}\mathbf{x}.$$

We will estimate the integral by integrating over the cross sections of the function in the integral. For the function values $t$, $0 \le t \le g_\delta^k(\delta^\alpha)$, all cross sections have the same form. For function values $t = g_\delta^k(y)$, $0 \le y < \delta^\alpha$, the cross section is $\{\mathbf{x} \in \mathbb{R}^d \mid d(\mathbf{x}, \partial\Omega) \le y\}$. We denote the area of the cross sections by $A(t)$. Integration of the areas gives

$$\begin{aligned}\int_{\{d(\mathbf{x},\partial\Omega)<\delta^\alpha\}} g_\delta^k(d(\mathbf{x},\partial\Omega))\mathrm{d}\mathbf{x} &= \int_0^{g_\delta^k(\delta^\alpha)} A(t)\mathrm{d}t + \int_{g_\delta^k(\delta^\alpha)}^{g_\delta^k(0)} A(t)\mathrm{d}t \\ &= A(g_\delta^k(\delta^\alpha))g_\delta^k(\delta^\alpha) + \int_{g_\delta^k(\delta^\alpha)}^{g_\delta^k(0)} A(t)\mathrm{d}t.\end{aligned}$$

We will use now the estimate of the areas of the cross sections given in Lemma 4.1. If $y$ is small enough, the term $y^d$ can be absorbed into the term $y$ in this estimate. Thus, if $\delta$ is small enough, we have $|\{\mathbf{x} \in \mathbb{R}^d \mid d(\mathbf{x}, \partial\Omega) \le y\}| \le Cy$, $0 \le y < \delta^\alpha$. We obtain, changing variables and applying integration by parts

$$\int_{g_\delta^k(\delta^\alpha)}^{g_\delta^k(0)} A(t)\mathrm{d}t \le -C\int_0^{\delta^\alpha} y\frac{\mathrm{d}}{\mathrm{d}y} g_\delta^k(y)\mathrm{d}y = C\left(-\delta^\alpha g_\delta^k(\delta^\alpha) + \int_0^{\delta^\alpha} g_\delta^k(y)\mathrm{d}y\right).$$

The last integral can be estimated further with the substitution $y = \delta s$:

$$\int_0^{\delta^\alpha} g_\delta^k(y)\mathrm{d}y \le C\delta^{1-kd}, \tag{20}$$

with $C$ depending on $k$. Collecting estimates, using $A(g_\delta(\delta^\alpha)) \le C\delta^\alpha$, which results in the cancellation of the terms $\delta^\alpha g_\delta^k(\delta^\alpha)$, we obtain

$$\int_{\mathbb{R}^d} B_\delta^k(\mathbf{x})\mathrm{d}\mathbf{x} \le C\delta^{1-kd+\frac{(d-1)\alpha k}{q}}.$$

The estimate for $C_\delta(\mathbf{x})$ converges exponentially for $\delta \to 0$. Thus, for $\delta$ sufficiently small, the estimate of $B_\delta(\mathbf{x})$ will dominate. This proves the proposition. □

## 5. Error estimates in the $(H^{-1}(\Omega))^d$–norm of the commutation error term

The main result of this section is that the commutation error tends to zero in $H^{-1}(\Omega)$ as $\delta \to 0$, see Proposition 5.1. The order of convergence is at least $O(\delta^{1/2})$.

**Lemma 5.1.** *There exists a constant $C$, which depends only on $d$, such that*

$$\|\overline{v} - v\|_{H^{1/2}(\mathbb{R}^d)} \le C\delta^{1/2}\|v\|_{H^1(\mathbb{R}^d)} \tag{21}$$

*for any $v \in H^1(\mathbb{R}^d)$ and any $\delta > 0$.*

*Proof.* By using the definition of $\| \cdot \|_{H^{1/2}(\mathbb{R}^d)}$, we have

$$\begin{aligned}\|\overline{v} - v\|^2_{H^{1/2}(\mathbb{R}^d)} &= \int_{\mathbb{R}^d} (1 + \|\mathbf{x}\|_2^2)^{1/2} |1 - \widehat{g}_\delta|^2 |\widehat{v}|^2 \mathbf{dx} \\ &= \int_{\{\|\mathbf{x}\|_2 > \pi/\delta\}} (1 + \|\mathbf{x}\|_2^2)^{1/2} |1 - \widehat{g}_\delta|^2 |\widehat{v}|^2 \mathbf{dx} \\ &\quad + \int_{\{\|\mathbf{x}\|_2 \le \pi/\delta\}} (1 + \|\mathbf{x}\|_2^2)^{1/2} |1 - \widehat{g}_\delta|^2 |\widehat{v}|^2 \mathbf{dx},\end{aligned}$$

where $\widehat{v}$ denotes the Fourier transform of $v$ and the Fourier transform of the Gaussian filter is given by

$$\widehat{g}_\delta(\mathbf{x}) = \exp\left(-\frac{\delta^2}{24}\|\mathbf{x}\|_2^2\right). \tag{22}$$

First, we prove a bound for the first integral. There exists a constant $C > 0$, which does not depend on $\delta$ and $v$, such that $(1 + \|\mathbf{x}\|_2^2)^{-1/2} < C\delta$ for $\|\mathbf{x}\|_2 > \pi/\delta$. From (22) follows the pointwise estimate $|1 - \widehat{g}_\delta(\mathbf{x})| \le 1$ for any $\mathbf{x} \in \mathbb{R}^d$. Thus, the first integral can be bounded by

$$\begin{aligned}&\left|\int_{\{\|\mathbf{x}\|_2 > \pi/\delta\}} (1 + \|\mathbf{x}\|_2^2)^{1/2} |1 - \widehat{g}_\delta|^2 |\widehat{v}|^2 \mathbf{dx}\right| \\ &\le \int_{\{\|\mathbf{x}\|_2 > \pi/\delta\}} (1 + \|\mathbf{x}\|_2^2)(1 + \|\mathbf{x}\|_2^2)^{-1/2} |\widehat{v}|^2 \mathbf{dx} \\ &\le C\delta \int_{\{\|\mathbf{x}\|_2 > \pi/\delta\}} (1 + \|\mathbf{x}\|_2^2) |\widehat{v}|^2 \mathbf{dx}.\end{aligned} \tag{23}$$

A Taylor series expansion of (22) at $\|\mathbf{x}\|_2 = 0$ and for fixed $\delta$ gives

$$\widehat{g}_\delta(\mathbf{x}) = 1 - \frac{\delta^2 \|\mathbf{x}\|_2^2}{24} + O(\delta^4 \|\mathbf{x}\|_2^4),$$

such that we have the pointwise bound

$$|1 - \widehat{g}_\delta(\mathbf{x})|^2 \le C\delta \|\mathbf{x}\|_2$$

for any $\|\mathbf{x}\|_2 \le \pi/\delta$ where C does not depend on $\delta$ or $\mathbf{x}$. In addition, $\|\mathbf{x}\|_2 \le (1 + \|\mathbf{x}\|_2^2)^{1/2}$ and consequently the second integral can be bounded as follows:

$$\left|\int_{\{\|\mathbf{x}\|_2 \le \pi/\delta\}} (1 + \|\mathbf{x}\|_2^2)^{1/2} |1 - \widehat{g}_\delta|^2 |\widehat{v}|^2 \mathbf{dx}\right| \le C\delta \int_{\{\|\mathbf{x}\|_2 \le \pi/\delta\}} (1 + \|\mathbf{x}\|_2^2) |\widehat{v}|^2 \mathbf{dx}. \tag{24}$$

Combining (23) and (24) gives

$$\|\overline{v} - v\|^2_{H^{1/2}(\mathbb{R}^d)} \le C\delta \int_{\mathbb{R}^d} (1 + \|\mathbf{x}\|_2^2) |\widehat{v}|^2 \mathbf{dx} = C\delta \|v\|_{H^1(\mathbb{R}^d)}. \qquad \square$$

Let $H^{-1}(\Omega)$ be the dual space of $H_0^1(\Omega)$ equipped with the norm

$$\|w\|_{H^{-1}(\Omega)} = \sup_{v \in H_0^1(\Omega)} \frac{\langle v, w\rangle}{\|v\|_{H^1(\Omega)}},$$

where $\langle\cdot,\cdot\rangle$ denotes the dual pairing.

**Proposition 5.1.** *Let $\psi \in L^2(\partial\Omega)$, then there exists a constant $C > 0$ which depends only on $\Omega$ such that*

$$\left\| \int_{\partial\Omega} g_\delta(\mathbf{x}-\mathbf{s})\psi(\mathbf{s})d\mathbf{s} \right\|_{H^{-1}(\Omega)} \le C\delta^{1/2}\|\psi\|_{L^2(\partial\Omega)}$$

*for every $\delta > 0$.*

*Proof.* Let $v \in H_0^1(\Omega)$. Extending $v$ by zero outside $\Omega$, applying Fubini's theorem, using that $v$ vanishes on $\partial\Omega$, applying the Cauchy–Schwarz inequality, the trace theorem and Lemma 5.1, give

$$\begin{aligned}
\int_\Omega \left( \int_{\partial\Omega} g_\delta(\mathbf{x}-\mathbf{s})\psi(\mathbf{s})d\mathbf{s} \right) v(\mathbf{x})\, d\mathbf{x} &= \int_{\partial\Omega} \psi(\mathbf{s})\overline{v}(\mathbf{s})ds \\
&= \int_{\partial\Omega} \psi(\mathbf{s})\left(\overline{v}(\mathbf{s}) - v(\mathbf{s})\right) ds \\
&\le \|\overline{v} - v\|_{L^2(\partial\Omega)}\|\psi\|_{L^2(\partial\Omega)} \\
&\le C\|\overline{v} - v\|_{H^{1/2}(\Omega)}\|\psi\|_{L^2(\partial\Omega)} \\
&\le C\delta^{1/2}\|v\|_{H^1(\Omega)}\|\psi\|_{L^2(\partial\Omega)}.
\end{aligned}$$

Division by $\|v\|_{H^1(\Omega)}$ and using the definition of the $H^{-1}(\Omega)$ norm gives the desired result. □

Let

$$\mathcal{H} = \left\{ v \in H^1(\mathbb{R}^d) \ : \ v|_{\partial\Omega} = 0 \right\}$$

and let the assumption of Proposition 5.1 be fulfilled. An inspection of the proof shows that then also

$$\begin{aligned}
\left\| \int_{\partial\Omega} g_\delta(\mathbf{x}-\mathbf{s})\psi(\mathbf{s})d\mathbf{s} \right\|_{H_{\mathcal{H}}^{-1}(\mathbb{R}^d)} &\le \sup_{v \in \mathcal{H}} \frac{\left\langle v, \int_{\partial\Omega} g_\delta(\mathbf{x}-\mathbf{s})\psi(\mathbf{s})d\mathbf{s} \right\rangle}{\|v\|_{H^1(\mathbb{R}^d)}} \\
&\le C\delta^{1/2}\|\psi\|_{L^2(\partial\Omega)}.
\end{aligned}$$

## 6. Error estimates for a weak form of the commutation error term

In this section, we consider a weak form of the commutation error term, $A_\delta(\mathbb{S}(\mathbf{u}, p))$, multiplied with a suitable test function $\overline{v}(\mathbf{x})$ and integrated on $\mathbb{R}^d$. The following proposition shows that this weak form converges to zero as $\delta$ tends to zero for fixed $\overline{v}(\mathbf{x})$. For $d = 2$, Corollary 6.1 and Remark 6.1 show that the convergence is (at least) almost of order one if $\psi$ is sufficiently smooth.

**Lemma 6.1.** *Let $v \in H^1(\mathbb{R}^d)$ such that $v|_\Omega \in H_0^1(\Omega) \cap H^2(\Omega)$ and $v(\mathbf{x}) = 0$ if $\mathbf{x} \notin \overline{\Omega}$, and let $\psi \in L^p(\partial\Omega)$, $1 \le p \le \infty$. Then*

$$\lim_{\delta\to 0} \int_{\mathbb{R}^d} \overline{v}(\mathbf{x}) \left( \int_{\partial\Omega} g_\delta(\mathbf{x}-\mathbf{s})\psi(\mathbf{s})d\mathbf{s} \right) d\mathbf{x} = 0,$$

*where $\overline{v}(\mathbf{x}) = (g_\delta * v)(\mathbf{x})$.*

*Proof.* By Fubini's theorem and the symmetry of $g_\delta$, we obtain

$$\begin{aligned} &\lim_{\delta\to 0} \int_{\mathbb{R}^d} \overline{v}(\mathbf{x}) \left( \int_{\partial\Omega} g_\delta(\mathbf{x}-\mathbf{s})\psi(\mathbf{s})d\mathbf{s} \right) d\mathbf{x} \\ &= \lim_{\delta\to 0} \int_{\partial\Omega} \psi(\mathbf{s}) \left( \int_{\mathbb{R}^d} g_\delta(\mathbf{s}-\mathbf{x})\overline{v}(\mathbf{x})dx \right) d\mathbf{s}. \end{aligned}$$

By a Sobolev imbedding theorem, we get $v \in L^\infty(\Omega)$ from what follows by the construction of $v$ that $v \in L^\infty(\mathbb{R}^d)$. Hölder's inequality for convolutions gives $\overline{v} \in L^\infty(\mathbb{R}^d)$. In addition, $\overline{v}$ is uniformly continuous on the compact set $\partial\Omega$. The same holds for $v$ since $v \in C^0(\overline{\Omega})$ by a Sobolev imbedding theorem. Applying twice Lemma 3.1 gives

$$\lim_{\delta\to 0} \int_{\mathbb{R}^d} \overline{v}(\mathbf{x}) \left( \int_{\partial\Omega} g_\delta(\mathbf{x}-\mathbf{s})\psi(\mathbf{s})d\mathbf{s} \right) d\mathbf{x} = \int_{\partial\Omega} \psi(\mathbf{s})v(\mathbf{s})d\mathbf{s} = 0,$$

since $v$ vanishes on $\partial\Omega$. □

With the result of Proposition 4.3, we want to study the order of convergence with respect to $\delta$ of the weak form of the commutation error term.

**Proposition 6.1.** *Let $v$ and $\psi$ be defined as in Lemma 6.1 and let the assumption of Proposition 4.3 be fulfilled. Then, there exists an $\epsilon > 0$ such that for $\delta \in (0, \epsilon)$*

$$\int_{\mathbb{R}^d} \left| \overline{v}(\mathbf{x}) \int_{\partial\Omega} g_\delta(\mathbf{x}-\mathbf{s})\psi(\mathbf{s})d\mathbf{s} \right|^k d\mathbf{x} \le C\delta^{1+(-d+\frac{(d-1)\alpha}{q}+\beta\alpha)k} \|\psi\|^k_{L^p(\partial\Omega)} \|v\|^k_{H^2(\Omega)},$$

*where $k \in [1, \infty)$, $\beta \in (0, 1)$ if $d = 2$ and $\beta = 1/2$ if $d = 3$, $p^{-1} + q^{-1} = 1$, $p > 1$, and $C$ and $\epsilon$ depend on $\alpha, k$ and $|\partial\Omega|$.*

*Proof.* Analogously to the begin of the proof of Proposition 4.3, one obtains

$$\begin{aligned} &\int_{\mathbb{R}^d} \left| \overline{v}(\mathbf{x}) \int_{\partial\Omega} g_\delta(\mathbf{x}-\mathbf{s})\psi(\mathbf{s})d\mathbf{s} \right|^k d\mathbf{x} \\ &\le C(k) \left[ \int_{\mathbb{R}^d} |\overline{v}(\mathbf{x})B_\delta(\mathbf{x})|^k \, d\mathbf{x} + \int_{\mathbb{R}^d} |\overline{v}(\mathbf{x})C_\delta(\mathbf{x})|^k \, d\mathbf{x} \right] \|\psi\|^k_{L^p(\partial\Omega)}, \end{aligned}$$

where $B_\delta(\mathbf{x})$ and $C_\delta(\mathbf{x})$ are defined in the proof of Proposition 4.3. The terms on the right hand side are treated separately.

In Proposition 4.3, it is proven that $C_\delta^k \in L^1(\mathbb{R}^d)$ for every $k \in (0, \infty)$. This implies

$$\left(C_\delta^k\right)^p = C_\delta^{kp} = C_\delta^{k'} \in L^1(\mathbb{R}^d),$$

since $k' \in (0,\infty)$. That means $C_\delta^k \in L^p(\mathbb{R}^d)$ for $p \in [1,\infty)$. From the bounding function of $C_\delta^k$ it is obvious that $C_\delta^k \in L^\infty(\mathbb{R}^d)$, too. Using Hölder's inequality for convolutions, see Adams [1, Theorem 4.30], and $\|g_\delta\|_{L^1(\mathbb{R}^d)} = 1$, it follows

$$\|\overline{v}\|_{L^q(\mathbb{R}^d)} \le \|g_\delta\|_{L^1(\mathbb{R}^d)}\|v\|_{L^q(\mathbb{R}^d)} = \|v\|_{L^q(\mathbb{R}^d)}$$

where $1 \le q < \infty$. With the same argument, we get for $qk \ge 1$

$$\|\overline{v}^k\|_{L^q(\mathbb{R}^d)} = \|\overline{v}\|^k_{L^{qk}(\mathbb{R}^d)} \le \|v\|^k_{L^{qk}(\mathbb{R}^d)}.$$

By the regularity assumptions on $v$, it follows $v \in C^0(\mathbb{R}^d)$. This implies, together with $v = 0$ outside $\Omega$, that $v \in L^p(\mathbb{R}^d)$ for every $1 \le p \le \infty$. Consequently, $\|v\|_{L^{qk}(\mathbb{R}^d)} < \infty$. Applying Hölder's inequality, we obtain

$$\int_{\mathbb{R}^d} |\overline{v}(\mathbf{x})C_\delta(\mathbf{x})|^k d\mathbf{x} \le \|v\|^k_{L^{qk}(\mathbb{R}^d)}\|C_\delta^k(\mathbf{x})\|_{L^p(\mathbb{R}^d)}.$$

For the second factor, we can use the bound obtained in the proof of Proposition 4.3, replacing $k$ by $kp$. Thus if $\delta$ is small enough, we obtain

$$\int_{\mathbb{R}^d} |\overline{v}(\mathbf{x})C_\delta(\mathbf{x})|^k d\mathbf{x} \le C\delta^{-kd}\left(\delta^{d\alpha}+\delta^\alpha\right)^{1/p}\exp\left(-\frac{6k}{\delta^{2(1-\alpha)}}\right)\|v\|^k_{L^{qk}(\mathbb{R}^d)} \qquad (25)$$

for every test function $v$ which satisfies the regularity assumptions stated in Lemma 6.1.

The estimate of the second term starts by noting that the domain of integration can be restricted to a small neighbourhood of $\partial\Omega$

$$\begin{aligned}\int_{\mathbb{R}^d} |\overline{v}(\mathbf{x})B_\delta(\mathbf{x})|^k d\mathbf{x} &= \int_{\{d(\mathbf{x},\partial\Omega)\le\delta^\alpha\}} |\overline{v}(\mathbf{x})B_\delta(\mathbf{x})|^k d\mathbf{x} \\ &\le \|\overline{v}\|^k_{L^\infty(\{d(\mathbf{x},\partial\Omega)\le\delta^\alpha\})}\int_{\{d(\mathbf{x},\partial\Omega)\le\delta^\alpha\}} B_\delta^k(\mathbf{x})d\mathbf{x} \\ &\le \|\overline{v}\|^k_{L^\infty(\{d(\mathbf{x},\partial\Omega)\le\delta^\alpha\})}\delta^{1+(-d+\frac{(d-1)\alpha}{q})k}, \qquad (26)\end{aligned}$$

where $\alpha \in (0,1)$ and $p^{-1}+q^{-1} = 1$. The last estimate is taken from the proof of Proposition 4.3. It remains to estimate the norm of $\overline{v}$. By the triangle inequality, we obtain

$$\|\overline{v}\|_{L^\infty(\{d(\mathbf{x},\partial\Omega)\le\delta^\alpha\})} \le \|\overline{v}-v\|_{L^\infty(\{d(\mathbf{x},\partial\Omega)\le\delta^\alpha\})} + \|v\|_{L^\infty(\{d(\mathbf{x},\partial\Omega)\le\delta^\alpha\})}. \qquad (27)$$

Since $v \in H^2(\Omega)$, we have $v \in C^{0,\beta}(\overline{\Omega})$ with $\beta \in (0,1)$ if $d = 2$ and $\beta = 1/2$ if $d = 3$. That means, there exists a constant $C_H \ge 0$ such that

$$|v(\mathbf{x}) - v(\mathbf{y})| \le C_H\|\mathbf{x}-\mathbf{y}\|_2^\beta \text{ for all } \mathbf{x},\mathbf{y} \in \overline{\Omega}.$$

By the Sobolev imbedding theorem, this constant can be estimated by $C_H \le C(\Omega)\|v\|_{H^2(\Omega)}$. We fix an arbitrary $\mathbf{x} \in \{d(\mathbf{x},\partial\Omega) \le \delta^\alpha\}$ and we take $\mathbf{y} \in \partial\Omega$ with $\|\mathbf{x}-\mathbf{y}\|_2 = d(\mathbf{x},\mathbf{y})$. Since $v$ vanishes on $\partial\Omega$, we obtain $\|v(\mathbf{x})\|_2 \le C_H d(\mathbf{x},\partial\Omega)^\beta$. It follows

$$\|v\|_{L^\infty(\{d(\mathbf{x},\partial\Omega)\le\delta^\alpha\})} \le C_H\delta^{\alpha\beta}.$$

The first term on the right hand side of (27) is, using that the $L^1(\mathbb{R}^d)$ norm of the Gaussian filter is equal to one,

$$\|\overline{v} - v\|_{L^\infty(\{d(\mathbf{x},\partial\Omega)\le\delta^\alpha\})} = \text{ess} \sup_{\mathbf{x}\in\{d(\mathbf{x},\partial\Omega)\le\delta^\alpha\}} \left|\int_{\mathbb{R}^d} g_\delta(\mathbf{x}-\mathbf{y})(v(\mathbf{x})-v(\mathbf{y}))d\mathbf{y}\right|.$$

Since $v$ vanishes outside $\Omega$, it can be easily proven that

$$|v(\mathbf{x}) - v(\mathbf{y})| \le C_H\|\mathbf{x}-\mathbf{y}\|_2^\beta$$

holds for all $\mathbf{x},\mathbf{y}\in\mathbb{R}^d$. We obtain, using the symmetry of the Gaussian filter,

$$\begin{aligned}
&\left|\int_{\mathbb{R}^d} g_\delta(\mathbf{x}-\mathbf{y})(v(\mathbf{x})-v(\mathbf{y}))d\mathbf{y}\right| \\
&\le C_H\int_{\mathbb{R}^d} g_\delta(\mathbf{x}-\mathbf{y})\|\mathbf{x}-\mathbf{y}\|_2^\beta d\mathbf{y} = C_H\int_{\mathbb{R}^d} g_\delta(\delta\mathbf{z})\delta^{\beta+d}\|\mathbf{z}\|_2^\beta d\mathbf{z} \\
&= CC_H\delta^\beta\int_{\mathbb{R}^d}\exp(-6\|\mathbf{z}\|_2^2)\|\mathbf{z}\|_2^\beta d\mathbf{z}.
\end{aligned}$$

The last integral is finite. Thus, we can conclude

$$\|\overline{v} - v\|_{L^\infty(\{d(x,\partial\Omega)\le\delta^\alpha\})} \le CC_H\delta^\beta.$$

Since $\delta^\beta$ decays faster for small $\delta$ than $\delta^{\beta\alpha}$, we obtain the estimate

$$\|\overline{v}\|^k_{L^\infty(\{d(\mathbf{x},\partial\Omega)\le\delta^\alpha\})} \le C_H^k\delta^{\beta\alpha k}.$$

Combining this estimate with (26) and using the estimate for $C_H$, we get

$$\int_{\mathbb{R}^d}|\overline{v}(\mathbf{x})B_\delta(\mathbf{x})|^k d\mathbf{x} \le C\delta^{1+(-d+\frac{(d-1)\alpha}{q}+\beta\alpha)k}\|v\|^k_{H^2(\Omega)}.$$

This dominates estimate (25) for small $\delta$. Collecting terms, gives the final result. □

An easy consequence of Proposition 6.1 is the following

**Corollary 6.1.** *Let the assumptions of Proposition* 6.1 *be fulfilled. Then, the weak form of the commutation error is bounded :*

$$\left|\int_{\mathbb{R}^d}\overline{v}(\mathbf{x})\int_{\partial\Omega} g_\delta(\mathbf{x}-\mathbf{s})\psi(\mathbf{s})d\mathbf{s}\right| d\mathbf{x} \le C\delta^{1-d+\frac{(d-1)\alpha}{q}+\beta\alpha}\|\psi\|_{L^p(\partial\Omega)}\|v\|_{H^2(\Omega)}. \quad (28)$$

**Remark 6.1.** Let $d=2$ and $p<\infty$ arbitrary large. Then $q$ is arbitrary close to one. Choosing $\alpha$ and $\beta$ also arbitrary close to one leads to the following power of $\delta$ in (28):

$$1 + (-2 + (1-\epsilon_1) + (1-\epsilon_2)) = 1 - (\epsilon_1+\epsilon_2) = 1-\epsilon_3$$

for arbitrary small $\epsilon_1,\epsilon_2,\epsilon_3 > 0$. In this case, the convergence is almost of first order.

The result of Proposition 6.1 does not provide an order of convergence for $d=3$. Lemma 2.1 suggests choosing $p=4$, i.e. $q=4/3$. Then, the power of $\delta$ in (28) becomes $2(\alpha-1)$, which is negative for $\alpha<1$.

## 7. The boundedness of the kinetic energy for $\overline{\mathbf{u}}$ in some LES models

The space averaged Navier–Stokes equations

$$\begin{aligned}
\overline{\mathbf{u}}_t - \nu\Delta\overline{\mathbf{u}} + \nabla\cdot(\overline{\mathbf{u}\mathbf{u}^T}) + \nabla\overline{p} & & (29)\\
= \overline{\mathbf{f}} + \int_{\partial\Omega} g(\mathbf{x}-\mathbf{s})\,[\nu\nabla\mathbf{u}(t,\mathbf{s})\mathbf{n}(\mathbf{s}) - p(t,\mathbf{s})\mathbf{n}(\mathbf{s})]\,d\mathbf{s} \text{ in } (0,T)\times\mathbb{R}^d & &\\
\nabla\cdot\overline{\mathbf{u}} = 0 \qquad \text{in } (0,T)\times\mathbb{R}^d & & (30)\\
\overline{\mathbf{u}}\,|_{t=0} = \overline{\mathbf{u}}_0 \qquad \text{in } \mathbb{R}^d & & (31)
\end{aligned}$$

are not yet a closed system since the tensor $\overline{\mathbf{u}\mathbf{u}^T}$ is a priori not related to $\overline{\mathbf{u}}$ and $\overline{p}$. One central issue in LES is closure : modeling this tensor in terms of $\overline{\mathbf{u}}$ and $\overline{p}$.

In this section, we will apply the results of Section 6 to show that the kinetic energy of $\overline{\mathbf{u}}$ is bounded for a number of LES models including the previously omitted $A_\delta(\mathbb{S}(\mathbf{u},p))$ commutation error term if $\delta$ is sufficiently small.

### 7.1. The Smagorinsky model

We consider first one of the simplest LES model which goes back to Smagorinsky [24]. This model is obtained by

$$\overline{\mathbf{u}\mathbf{u}^T} \approx \overline{\mathbf{u}}\,\overline{\mathbf{u}}^T - C_\nu\delta^2\|\nabla\overline{\mathbf{u}}\|_F\nabla\overline{\mathbf{u}} + \text{terms absorbed into } \nabla\overline{p},$$

where $C_\nu > 0$ and $\|\nabla\overline{\mathbf{u}}\|_F = (\nabla\overline{\mathbf{u}} : \nabla\overline{\mathbf{u}})^{1/2}$ is the Frobenius norm of $\nabla\overline{\mathbf{u}}$.

The existence and uniqueness of a weak solution of the Smagorinsky model posed in a bounded domain, with homogeneous Dirichlet boundary conditions and without the boundary integral in (29) has been proven by Ladyzhenskaya [17, 18] and Du and Gunzburger [7].

The momentum equation of the Smagorinsky model has the form

$$\begin{aligned}
&\overline{\mathbf{u}}_t - \nabla\cdot\left(\left(\nu + C_\nu\delta^2\|\nabla\overline{\mathbf{u}}\|_F\right)\nabla\overline{\mathbf{u}}\right) + \nabla\cdot(\overline{\mathbf{u}}\,\overline{\mathbf{u}}^T) + \nabla\overline{p}\\
&= \overline{\mathbf{f}} + \int_{\partial\Omega} g_\delta(\mathbf{x}-\mathbf{s})\psi(t,\mathbf{s})d\mathbf{s} \qquad \text{in } (0,T)\times\mathbb{R}^d
\end{aligned} \tag{32}$$

with $\psi(t,\mathbf{s}) = \nu\nabla\mathbf{u}(t,\mathbf{s})\mathbf{n}(\mathbf{s}) - p(t,\mathbf{s})\mathbf{n}(\mathbf{s})$. We assume, there exists a solution $(\mathbf{u},p)$ of (32), (30), (31) Multiplying (32) by $\overline{\mathbf{u}}$ and integrating on $\mathbb{R}^d$ gives

$$\begin{aligned}
&\frac{\partial}{\partial t}\frac{\|\overline{\mathbf{u}}\|^2_{(L^2(\mathbb{R}^d))^d}}{2} - \int_{\mathbb{R}^d}\nabla\cdot\left(\left(\nu + C_\nu\delta^2\|\nabla\overline{\mathbf{u}}\|_F\right)\nabla\overline{\mathbf{u}}\right)\cdot\overline{\mathbf{u}}d\mathbf{x}\\
&\quad + \int_{\mathbb{R}^d}\nabla\cdot(\overline{\mathbf{u}}\,\overline{\mathbf{u}}^T)\cdot\overline{\mathbf{u}}d\mathbf{x} + \int_{\mathbb{R}^d}\nabla\overline{p}\cdot\overline{\mathbf{u}}d\mathbf{x}\\
&= \int_{\mathbb{R}^d}\overline{\mathbf{f}}\cdot\overline{\mathbf{u}}d\mathbf{x} + \int_{\mathbb{R}^d}\overline{\mathbf{u}}(t,\mathbf{x})\cdot\left(\int_{\partial\Omega} g_\delta(\mathbf{x}-\mathbf{s})\psi(t,\mathbf{s})d\mathbf{s}\right)d\mathbf{x}.
\end{aligned} \tag{33}$$

Next, the terms on the left hand side are studied. Using the definition of $\overline{\mathbf{u}}$, changing variables, noting that $\mathbf{u}$ vanishes outside $\Omega$ and applying Fubini's theorem, we obtain

$$\begin{aligned}
&\int_{\mathbb{R}^d} \nabla\overline{p}(\mathbf{x})\cdot\overline{\mathbf{u}}(\mathbf{x})d\mathbf{x}\\
&= \int_{\mathbb{R}^d} \mathbf{u}(\mathbf{y})\cdot\left(\int_{\mathbb{R}^d}\nabla\overline{p}(\mathbf{y}+\mathbf{z})g_\delta(\mathbf{z})d\mathbf{z}\right)d\mathbf{y}\\
&= \int_{\mathbb{R}^d} g_\delta(\mathbf{z})\left(\int_{\Omega}\nabla\overline{p}(\mathbf{y}+\mathbf{z})\cdot\mathbf{u}(\mathbf{y})d\mathbf{y}\right)d\mathbf{z}\\
&= \int_{\mathbb{R}^d} g_\delta(\mathbf{z})\left(-\int_{\Omega}\overline{p}(\mathbf{y}+\mathbf{z})(\nabla\cdot\mathbf{u})(\mathbf{y})d\mathbf{y}+\int_{\partial\Omega}\overline{p}(\mathbf{s}+\mathbf{z})\mathbf{u}(\mathbf{s})\cdot\mathbf{n}(\mathbf{s})d\mathbf{s}\right)d\mathbf{z}\\
&= 0,
\end{aligned}$$

since $\mathbf{u}$ is divergence free and vanishes on $\partial\Omega$. With an index calculation, using that $\overline{\mathbf{u}}$ is divergence free in $\mathbb{R}^d$, and applying the same arguments as for the pressure term, one obtains

$$\int_{\mathbb{R}^d}\nabla\cdot(\overline{\mathbf{u}}\,\overline{\mathbf{u}}^T)\cdot\overline{\mathbf{u}}d\mathbf{x} = \frac{1}{2}\int_{\mathbb{R}^d}\nabla(\overline{\mathbf{u}}\cdot\overline{\mathbf{u}})\cdot\overline{\mathbf{u}}d\mathbf{x} = 0.$$

By applying Fubini's theorem in the same way as before, it follows that

$$\begin{aligned}
&-\int_{\mathbb{R}^d}\nabla\cdot((\nu+C_\nu\delta^2\|\nabla\overline{\mathbf{u}}\|_F)\nabla\overline{\mathbf{u}})\cdot\overline{\mathbf{u}}d\mathbf{x}\\
&= \int_{\mathbb{R}^d}((\nu+C_\nu\delta^2\|\nabla\overline{\mathbf{u}}\|_F)\nabla\overline{\mathbf{u}}):\nabla\overline{\mathbf{u}}d\mathbf{x}\\
&\geq 0.
\end{aligned}$$

The first term on the right hand side of (33) is estimated by the Cauchy–Schwarz inequality and Young's inequality

$$\int_{\mathbb{R}^d}\overline{\mathbf{f}}\cdot\overline{\mathbf{u}}d\mathbf{x} \leq \|\overline{\mathbf{f}}\|_{(L^2(\mathbb{R}^d))^d}\|\overline{\mathbf{u}}\|_{(L^2(\mathbb{R}^d))^d} \leq \frac{\|\overline{\mathbf{f}}\|^2_{(L^2(\mathbb{R}^d))^d}}{2}+\frac{\|\overline{\mathbf{u}}\|^2_{(L^2(\mathbb{R}^d))^d}}{2}.$$

We obtain

$$\begin{aligned}
&\frac{\partial}{\partial t}\frac{\|\overline{\mathbf{u}}\|^2_{(L^2(\mathbb{R}^d))^d}}{2}\\
&\leq \frac{\|\overline{\mathbf{f}}\|^2_{(L^2(\mathbb{R}^d))^d}}{2}+\frac{\|\overline{\mathbf{u}}\|^2_{(L^2(\mathbb{R}^d))^d}}{2}+\int_{\mathbb{R}^d}\overline{\mathbf{u}}(t,\mathbf{x})\cdot\left(\int_{\partial\Omega}g_\delta(\mathbf{x}-\mathbf{s})\psi(t,\mathbf{s})d\mathbf{s}\right)d\mathbf{x}.
\end{aligned}$$

Assuming $\mathbf{f} \in L^2(0,T;(L^2(\mathbb{R}^d))^d)$, so that Gronwall's lemma can be applied, gives

$$\begin{aligned} &\frac{\|\overline{\mathbf{u}}(T)\|^2_{(L^2(\mathbb{R}^d))^d}}{2} \\ &\leq \exp(T)\frac{\|\overline{\mathbf{u}}_0\|^2_{(L^2(\mathbb{R}^d))^d}}{2} + \int_0^T \exp(T-t)\frac{\|\overline{\mathbf{f}}(t)\|^2_{(L^2(\mathbb{R}^d))^d}}{2}dt \\ &\quad + \int_0^T \exp(T-t)\left(\int_{\mathbb{R}^d} \overline{\mathbf{u}}(t,\mathbf{x})\cdot\left(\int_{\partial\Omega} g_\delta(\mathbf{x}-\mathbf{s})\psi(t,\mathbf{s})d\mathbf{s}\right)d\mathbf{x}\right)dt. \end{aligned} \tag{34}$$

Thus, the kinetic energy of $\overline{\mathbf{u}}$ at time $T$ is bounded by the kinetic energy of the initial data, by the right hand side and by a third term, which vanishes as $\delta \to 0$ by Proposition 6.1. For $d = 2$, it follows from Remark 6.1, (5) and Young's inequality

$$\begin{aligned} &\frac{\|\overline{\mathbf{u}}(T)\|^2_{(L^2(\mathbb{R}^2))^2}}{2} \\ &\leq \exp(T)\frac{\|\overline{\mathbf{u}}_0\|^2_{(L^2(\mathbb{R}^2))^2}}{2} + \int_0^T \exp(T-t)\frac{\|\overline{\mathbf{f}}(t)\|^2_{(L^2(\mathbb{R}^2))^2}}{2}dt \\ &\quad + C\delta^{1-\epsilon_3}\int_0^T \exp(T-t)\left(\|\mathbf{u}(t)\|^2_{(H^2(\Omega))^2} + \|p(t)\|^2_{H^1(\Omega)}\right)dt \end{aligned} \tag{35}$$

with arbitrary $\epsilon_3 > 0$.

**Remark 7.1.** One obtains the same result for the deformation tensor formulation of the momentum equation and a Smagorinsky model of the form

$$\overline{\mathbf{u}\mathbf{u}^T} \approx \overline{\mathbf{u}}\,\overline{\mathbf{u}}^T - C_\nu\delta^2\|\mathbb{D}(\overline{\mathbf{u}})\|_F\mathbb{D}(\overline{\mathbf{u}}) + \text{terms absorbed into } \nabla\overline{p}.$$

### 7.2. The Taylor LES model

The second model which we will discuss is called variously gradient method of the Taylor LES model developed by Leonard [19] and Clark et al. [4]. In the mixed Taylor LES model, a Smagorinsky model for the turbulent fluctuations is included in the model of the non–linear convective term given by

$$\overline{\mathbf{u}\mathbf{u}^T} \approx \overline{\mathbf{u}}\,\overline{\mathbf{u}}^T - C_\nu\delta^2\|\nabla\overline{\mathbf{u}}\|_F\nabla\overline{\mathbf{u}} + \frac{\delta^2}{12}\nabla\overline{\mathbf{u}}\nabla\overline{\mathbf{u}}^T.$$

The existence and uniqueness of a weak solution of the Taylor LES model in a bounded domain, equipped with homogeneous Dirichlet boundary conditions and without the boundary integral in (29) has been proven for $C_\nu$ large enough by Coletti [5]. We study the energy balance of the Taylor LES model including the term $A_\delta(\mathbb{S}(\mathbf{u},p))$

Inserting the Taylor LES model into (29), multiplying by $\overline{\mathbf{u}}$ and integrating by parts, the non-linear convective term and the pressure term vanish as in the Smagorinsky model. The bilinear viscous term is obviously non-negative. The non-linear viscous term is treated in connection with the third term in the Taylor LES model. Using the same arguments as in the Smagorinsky model, one finds

$$\int_{\mathbb{R}^d} \nabla \cdot \left( -C_\nu \delta^2 \|\nabla \overline{\mathbf{u}}\|_F \nabla \overline{\mathbf{u}} + \frac{\delta^2}{12} \nabla \overline{\mathbf{u}} \nabla \overline{\mathbf{u}}^T \right) \cdot \overline{\mathbf{u}} d\mathbf{x}$$
$$= \int_{\mathbb{R}^d} C_\nu \delta^2 \|\nabla \overline{\mathbf{u}}\|_F \nabla \overline{\mathbf{u}} : \nabla \overline{\mathbf{u}} - \frac{\delta^2}{12} (\nabla \overline{\mathbf{u}} \nabla \overline{\mathbf{u}}^T) : \nabla \overline{\mathbf{u}} d\mathbf{x}. \tag{36}$$

By norm equivalence in $\mathbb{R}^d \times \mathbb{R}^d$,

$$\left( \sum_{i=1}^{d} |a_{ij}|^3 \right)^{\frac{1}{3}} \leq \|A\|_F \leq C(d) \left( \sum_{i=1}^{d} |a_{ij}|^3 \right)^{\frac{1}{3}},$$

we get

$$\int_{\mathbb{R}^d} C_\nu \delta^2 \|\nabla \overline{\mathbf{u}}\|_F \nabla \overline{\mathbf{u}} : \nabla \overline{\mathbf{u}} d\mathbf{x}$$
$$= \int_{\mathbb{R}^d} C_\nu \delta^2 \|\nabla \overline{\mathbf{u}}\|_F^3 d\mathbf{x} \geq C_\nu \delta^2 \|\nabla \overline{\mathbf{u}}\|_{L^3(\mathbb{R}^d)}^3.$$

The second term in (36) is estimated in a similar way. Using twice the Cauchy–Schwarz inequality, one obtains

$$\int_{\mathbb{R}^d} \frac{\delta^2}{12} (\nabla \overline{\mathbf{u}} \nabla \overline{\mathbf{u}}^T) : \nabla \overline{\mathbf{u}} d\mathbf{x} \leq \int_{\mathbb{R}^d} \frac{\delta^2}{12} \|\nabla \overline{\mathbf{u}}\|_F^3 d\mathbf{x} \leq C(d) \frac{\delta^2}{12} \|\nabla \overline{\mathbf{u}}\|_{L^3(\mathbb{R}^d)}^3.$$

We get finally

$$-\int_{\mathbb{R}^d} \left( -C_\nu \delta^2 \|\nabla \overline{\mathbf{u}}\|_F \nabla \overline{\mathbf{u}} + \frac{\delta^2}{12} \nabla \overline{\mathbf{u}} \nabla \overline{\mathbf{u}}^T \right) : \nabla \overline{\mathbf{u}} d\mathbf{x}$$
$$\geq C_\nu \delta^2 \|\nabla \overline{\mathbf{u}}\|_{L^3(\mathbb{R}^d)}^3 - C(d) \frac{\delta^2}{12} \|\nabla \overline{\mathbf{u}}\|_{L^3(\mathbb{R}^d)}^3 \geq 0$$

if $C_\nu$ is sufficiently large.

Now, we can continue as for the Smagorinsky model and obtain also the estimates (34) and (35) for the kinetic energy of $\overline{\mathbf{u}}$ if $C_\nu$ is chosen large enough and if $\delta$ is sufficiently small.

### 7.3. The rational LES model

The rational LES model was proposed in [11]. Including the Smagorinsky model for the effects of turbulent fluctuations, there are two variants of this model, one of them has the form

$$\overline{\mathbf{u}\mathbf{u}^T} \approx \overline{\mathbf{u}}\,\overline{\mathbf{u}}^T - C_\nu \delta^2 \|\nabla \overline{\mathbf{u}}\|_F \nabla \overline{\mathbf{u}} + \frac{\delta^2}{12} g_\delta * \left( \nabla \overline{\mathbf{u}} \nabla \overline{\mathbf{u}}^T \right).$$

The existence and uniqueness of generalized solutions of the rational LES model in the space periodic case for small data and small times has been proven by Berselli et al. [3].

Proceeding as in the Taylor LES model, the only difference is the term

$$\int_{\mathbb{R}^d} C_\nu \delta^2 \|\nabla \overline{\mathbf{u}}\|_F^3 - \frac{\delta^2}{12} g_\delta * \left( \nabla \overline{\mathbf{u}} \nabla \overline{\mathbf{u}}^T \right) : \nabla \overline{\mathbf{u}} d\mathbf{x}.$$

The application of Fubini's theorem and the symmetry of the Gaussian filter yield

$$\int_{\mathbb{R}^d} g_\delta * (\nabla\overline{\mathbf{u}}\nabla\overline{\mathbf{u}}^T) : \nabla\overline{\mathbf{u}}d\mathbf{x} = \int_{\mathbb{R}^d} \nabla\overline{\mathbf{u}}\nabla\overline{\mathbf{u}}^T : (g_\delta * \nabla\overline{\mathbf{u}})d\mathbf{x}.$$

It follows with the same arguments as in the estimate for the Taylor LES model, using in addition Hölder's inequality for convolutions,

$$\begin{aligned}\int_{\mathbb{R}^d} g_\delta * \left(\nabla\overline{\mathbf{u}}\nabla\overline{\mathbf{u}}^T\right) : \nabla\overline{\mathbf{u}}d\mathbf{x} &\le \int_{\mathbb{R}^d} \|\nabla\overline{\mathbf{u}}\|_F^2 \|g_\delta * \nabla\overline{\mathbf{u}}\|_F d\mathbf{x} \\ &\le \|\nabla\overline{\mathbf{u}}\|_{L^3(\mathbb{R}^d)}^2 \|g_\delta * \nabla\overline{\mathbf{u}}\|_{L^3(\mathbb{R}^d)} \\ &\le \|\nabla\overline{\mathbf{u}}\|_{L^3(\mathbb{R}^d)}^3 \|g_\delta\|_{L^1(\mathbb{R}^d)} = \|\nabla\overline{\mathbf{u}}\|_{L^3(\mathbb{R}^d)}^3.\end{aligned}$$

This gives the estimate

$$\begin{aligned}&\int_{\mathbb{R}^d} C_\nu\delta^2\|\nabla\overline{\mathbf{u}}\|_F^3 - \frac{\delta^2}{12} g_\delta * \left(\nabla\overline{\mathbf{u}}\nabla\overline{\mathbf{u}}^T\right) : \nabla\overline{\mathbf{u}}d\mathbf{x} \\ &\quad \ge \left(C_\nu\delta^2 - C_K^3 C(d)\frac{\delta^2}{12}\right)\|\nabla\overline{\mathbf{u}}\|_{L^3(\mathbb{R}^d)}^3 \ge 0\end{aligned}$$

if $C_\nu$ is large enough.

That means, also for the rational LES model, the kinetic energy of $\overline{\mathbf{u}}$ can be estimated in form (34) and (35) if $C_\nu$ is chosen sufficiently large and $\delta$ sufficiently small.

**Acknowledgment**

We thank Prof. G.P. Galdi for pointing out the result of Giga and Sohr in Remark 2.1.

## References

[1] R.A. Adams. *Sobolev spaces.* Academic Press, New York, 1975.

[2] A.A. Aldama. *Filtering Techniques for Turbulent Flow Simulation*, volume 56 of *Springer Lecture Notes in Eng.* Springer, Berlin, 1990.

[3] L.G. Berselli, G.P. Galdi, W.J. Layton, and T. Iliescu. Mathematical analysis for the rational large eddy simulation model. *Math. Models and Meth. in Appl. Sciences*, 12:1131–1152, 2002.

[4] R.A. Clark, J. H. Ferziger, and W.C. Reynolds. Evaluation of subgrid-scale models using an accurately simulated turbulent flow. *J. Fluid Mech.*, 91:1–16, 1979.

[5] P. Coletti. *Analytic and Numerical Results for $k - \varepsilon$ and Large Eddy Simulation Turbulence Models.* PhD thesis, University of Trento, 1998.

[6] A. Das and R.D. Moser. Filtering boundary conditions for LES and embedded boundary simulations. In C. Liu, L. Sakell, and T. Beutner, editors, *DNS/LES - Progress and Challenges (Proceedings of Third AFOSR International Conference on DNS and LES)*, pages 389–396. Greyden Press, Columbus, 2001.

[7] Q. Du and M.D. Gunzburger. Analysis of a Ladyzhenskaya model for incompressible viscous flow. *J. Math. Anal. Appl.*, 155:21–45, 1991.

[8] G.B. Folland. *Introduction to Partial Differential Equations*, volume 17 of *Mathematical Notes.* Princeton University Press, 2nd edition, 1995.

[9] C. Fureby and G. Tabor. Mathematical and physical constraints on large-eddy simulations. *Theoret. Comput. Fluid Dynamics*, 9:85–102, 1997.

[10] G.P. Galdi. *An Introduction to the Mathematical Theory of the Navier-Stokes Equations, Vol. I: Linearized Steady Problems*, volume 38 of *Springer Tracts in Natural Philosophy.* Springer, 1994.

[11] G.P. Galdi and W.J. Layton. Approximation of the larger eddies in fluid motion II: A model for space filtered flow. *Math. Models and Meth. in Appl. Sciences*, 10(3):343–350, 2000.

[12] S. Ghosal and P. Moin. The basic equations for large eddy simulation of turbulent flows in complex geometries. *Journal of Computational Physics*, 118:24–37, 1995.

[13] Y. Giga and H. Sohr. Abstract $L^p$ estimates for the Cauchy problem with applications to the Navier–Stokes equations in exterior domains. *J. Funct. Anal.*, 102:72–94, 1991.

[14] L. Hörmander. *The Analysis of Partial Differential Operators I.* Springer - Verlag, Berlin, ..., 2nd edition, 1990.

[15] T.J. Hughes, L. Mazzei, and K.E. Jansen. Large eddy simulation and the variational multiscale method. *Comput. Visual. Sci.*, 3:47–59, 2000.

[16] V. John and W.J. Layton. Approximating local averages of fluid velocities: The Stokes problem. *Computing*, 66:269–287, 2001.

[17] O.A. Ladyzhenskaya. New equations for the description of motion of viscous incompressible fluids and solvability in the large of boundary value problems for them. *Proc. Steklov Inst. Math.*, 102:95–118, 1967.

[18] O.A. Ladyzhenskaya. *The Mathematical Theory of Viscous Incompressible Flow.* Gordon and Breach, 2nd edition, 1969.

[19] A. Leonard. Energy cascade in large eddy simulation of turbulent fluid flows. *Adv. in Geophysics*, 18A:237–248, 1974.

[20] M. Lesieur. *Turbulence in Fluids*, volume 40 of *Fluid Mechanics and its Applications.* Kluwer Academic Publishers, 3rd edition, 1997.

[21] S. B. Pope. *Turbulent flows.* Cambridge University Press, 2000.

[22] W. Rudin. *Functional Analysis.* International Series in Pure and Applied Mathematics. McGraw-Hill, Inc., New York, 2nd edition, 1991.

[23] P. Sagaut. *Large Eddy Simulation for Incompressible Flows.* Springer–Verlag Berlin Heidelberg New York, 2001.

[24] J.S. Smagorinsky. General circulation experiments with the primitive equations. *Mon. Weather Review*, 91:99–164, 1963.

[25] O.V. Vasilyev, T.S. Lund, and P. Moin. A general class of commutative filters for LES in complex geometries. *Journal of Computational Physics*, 146:82–104, 1998.

A. Dunca
Department of Mathematics
University of Pittsburgh
Pittsburgh, PA 15260
U.S.A.
e-mail: `ardst21+@pitt.edu`

V. John
Institut für Analysis und Numerik
Otto-von-Guericke-Universität Magdeburg
PF 4120
D-39016 Magdeburg
Germany
e-mail: `john@mathematik.uni-magdeburg.de`
Homepage: `http://www-ian.math.uni-magdeburg.de/home/john/`

W.J. Layton
Department of Mathematics
University of Pittsburgh
Pittsburgh, PA 15260
U.S.A.
e-mail: `wjl@pitt.edu`
Homepage: `http: http://www.math.pitt.edu/~wjl;`

Advances in Mathematical Fluid Mechanics, 79–123


# The Nonstationary Stokes and Navier–Stokes Flows Through an Aperture

Toshiaki Hishida

**Abstract.** We consider the nonstationary Stokes and Navier–Stokes flows in aperture domains $\Omega \subset \boldsymbol{R}^n, n \geq 3$. We develop the $L^q$-$L^r$ estimates of the Stokes semigroup and apply them to the Navier–Stokes initial value problem. As a result, we obtain the global existence of a unique strong solution, which satisfies the vanishing flux condition through the aperture and some sharp decay properties as $t \to \infty$, when the initial velocity is sufficiently small in the $L^n$ space. Such a global existence theorem is up to now well known in the cases of the whole and half spaces, bounded and exterior domains.

**Mathematics Subject Classification (2000).** 35Q30, 76D05.

**Keywords.** Aperture domain, Navier–Stokes flow, Stokes semigroup, decay estimate.

*Dedicated to the memory of the late Professor Yasujiro Nagakura*

## 1. Introduction

In the present paper we study the global existence and asymptotic behavior of a strong solution to the Navier–Stokes initial value problem in an aperture domain $\Omega \subset \boldsymbol{R}^n$ with smooth boundary $\partial\Omega$:

$$\begin{cases} \partial_t u + u \cdot \nabla u &= \Delta u - \nabla p \qquad (x \in \Omega,\ t > 0), \\ \nabla \cdot u &= 0 \qquad (x \in \Omega,\ t \geq 0), \\ u|_{\partial\Omega} &= 0 \qquad (t > 0), \\ u|_{t=0} &= a \qquad (x \in \Omega), \end{cases} \tag{1.1}$$

where $u(x,t) = (u_1(x,t), \cdots, u_n(x,t))$ and $p(x,t)$ denote the unknown velocity and pressure of a viscous incompressible fluid occupying $\Omega$, respectively, while $a(x) = (a_1(x), \cdots, a_n(x))$ is a prescribed initial velocity. The aperture domain $\Omega$

---

On leave of absence from Niigata University, Niigata 950-2181, Japan (e-mail: hishidaeng.niigata-u.ac.jp). Supported in part by the Alexander von Humboldt research fellowship.

is a compact perturbation of two separated half spaces $H_+ \cup H_-$, where $H_\pm = \{x = (x_1, \cdots, x_n) \in \boldsymbol{R}^n; \pm x_n > 1\}$; to be precise, we call a connected open set $\Omega \subset \boldsymbol{R}^n$ an aperture domain (with thickness of the wall) if there is a ball $B \subset \boldsymbol{R}^n$ such that $\Omega \setminus B = (H_+ \cup H_-) \setminus B$. Thus the upper and lower half spaces $H_\pm$ are connected by an aperture (hole) $M \subset \Omega \cap B$, which is a smooth $(n-1)$-dimensional manifold so that $\Omega$ consists of upper and lower disjoint subdomains $\Omega_\pm$ and $M$: $\Omega = \Omega_+ \cup M \cup \Omega_-$.

The aperture domain is a particularly interesting class of domains with noncompact boundaries because of the following remarkable feature, which was in 1976 pointed out by Heywood [34]: the solution is not uniquely determined by usual boundary conditions even for the stationary Stokes system in this domain and therefore, in order to single out a unique solution, we have to prescribe either the flux through the aperture $M$

$$\phi(u) = \int_M N \cdot u d\sigma,$$

or the pressure drop at infinity (in a sense) between the upper and lower subdomains $\Omega_\pm$

$$[p] = \lim_{|x| \to \infty, x \in \Omega_+} p(x) - \lim_{|x| \to \infty, x \in \Omega_-} p(x),$$

as an additional boundary condition. Here, $N$ denotes the unit normal vector on $M$ directed to $\Omega_-$ and the flux $\phi(u)$ is independent of the choice of $M$ since $\nabla \cdot u = 0$ in $\Omega$. Consider stationary solutions of (1.1); then one can formally derive the energy relation

$$\int_\Omega |\nabla u(x)|^2 dx = [p]\phi(u),$$

from which the importance of these two physical quantities stems. Later on, the observation of Heywood in the $L^2$ framework was developed by Farwig and Sohr within the framework of $L^q$ theory for the stationary Stokes and Navier–Stokes systems [21] and also the (generalized) Stokes resolvent system [22], [18]. Especially, in the latter case, they clarified that the assertion on the uniqueness depends on the class of solutions under consideration. Indeed, the additional condition must be required for the uniqueness if $q > n/(n-1)$, but otherwise, the solution is unique without any additional condition; for more details, see Farwig [18, Theorem 1.2].

The results of Farwig and Sohr [22] are also the first step to discuss the nonstationary problem (1.1) in the $L^q$ space. They, as well as Miyakawa [53], showed the Helmholtz decomposition of the $L^q$ space of vector fields $L^q(\Omega) = L^q_\sigma(\Omega) \oplus L^q_\pi(\Omega)$ for $n \geq 2$ and $1 < q < \infty$, where $L^q_\sigma(\Omega)$ is the completion in $L^q(\Omega)$ of the class of all smooth, solenoidal and compactly supported vector fields, and $L^q_\pi(\Omega) = \{\nabla p \in L^q(\Omega); p \in L^q_{loc}(\overline{\Omega})\}$. The space $L^q_\sigma(\Omega)$ is characterized as ([22, Lemma 3.1], [53, Theorem 4])

$$L^q_\sigma(\Omega) = \{u \in L^q(\Omega); \nabla \cdot u = 0, \nu \cdot u|_{\partial\Omega} = 0, \phi(u) = 0\}, \tag{1.2}$$

where $\nu$ is the unit outer normal vector on $\partial\Omega$. Here, the condition $\phi(u) = 0$ follows from the other ones and may be omitted if $q \leq n/(n-1)$, but otherwise, the element of $L^q_\sigma(\Omega)$ must possess this additional property. Using the projection $P_q$ from $L^q(\Omega)$ onto $L^q_\sigma(\Omega)$ associated with the Helmholtz decomposition, we can define the Stokes operator $A = A_q = -P_q\Delta$ on $L^q_\sigma(\Omega)$ with a right domain as in section 2. Then the operator $-A$ generates a bounded analytic semigroup $e^{-tA}$ in each $L^q_\sigma(\Omega), 1 < q < \infty$, for $n \geq 2$ ([22, Theorem 2.5]).

Besides [34] and [21] cited above, there are some other studies on the stationary Stokes and Navier-Stokes systems in domains with noncompact boundaries including aperture domains. We refer to Borchers and Pileckas [8], Borchers, Galdi and Pileckas [3], Galdi [28], Pileckas [55] and the references therein.

We are interested in strong solutions to the nonstationay problem (1.1). However, there are no results on the global existence of such solutions in the $L^q$ framework unless $q = 2$, while a few local existence theorems are known. In the 3-dimensional case, Heywood [34], [35] first constructed a local solution to (1.1) with a prescribed either $\phi(u(t))$ or $[p(t)]$, which should satisfy some regularity assumptions with respect to the time variable, when $a \in H^2(\Omega)$ fulfills some compatibility conditions. Franzke [25] has recently developed the $L^q$ theory of local solutions via the approach of Giga and Miyakawa [32], which is traced back to Fujita and Kato [26], with use of fractional powers of the Stokes operator. When a suitable $\phi(u(t))$ is prescribed, his assumption on initial data is for instance that $a \in L^q(\Omega), q > n$, together with some compatibility conditions. The reason why the case $q = n$ is excluded is the lack of information about purely imaginary powers of the Stokes operator. In order to discuss also the case where $[p(t)]$ is prescribed, Franzke introduced another kind of Stokes operator associated with the pressure drop condition, which generates a bounded analytic semigroup on the space $\{u \in L^q(\Omega); \nabla \cdot u = 0, \nu \cdot u|_{\partial\Omega} = 0\}$ for $n \geq 3$ and $n/(n-1) < q < n$ (based on a resolvent estimate due to Farwig [18]). Because of this restriction on $q$, the $L^q$ theory with $q \geq n$ is not available under the pressure drop condition and thus one cannot avoid a regularity assumption to some extent on initial data.

It is possible to discuss the $L^2$ theory of global strong solutions for an arbitrary unbounded domain (with smooth boundary) in a unified way since the Stokes operator is a nonnegative selfadjoint one in $L^2_\sigma$; see Heywood [36] ($n = 3$), Kozono and Ogawa [44] ($n = 2$), [45] ($n = 3$) and Kozono and Sohr [47] ($n = 4, 5$). Especially, from the viewpoint of the class of initial data, optimal results were given by [44], [45] and [47]. In fact, they constructed a global solution with various decay properties for small $a \in D(A_2^{n/4-1/2})$ (when $n = 2$, the smallness is not necessary). Here, we should recall the continuous embedding relation $D(A_2^{n/4-1/2}) \subset L^n_\sigma$. For the aperture domain $\Omega$ their solutions $u(t)$ should satisfy the hidden flux condition $\phi(u(t)) = 0$ on account of $u(t) \in L^2_\sigma(\Omega)$ together with (1.2). We also refer to Solonnikov [62], [63], in which a theory of generalized solutions was developed for a large class of domains having outlets to infinity. In his Doktorschrift [24] Franzke studied, among others, the global existence of weak and strong solutions in a 3-dimensional aperture domain when either $\phi(u(t))$ or $[p(t)]$ is prescribed (the global existence of the former for $n \geq 2$ is covered by Masuda [51] when $\phi(u(t)) = 0$). As for the latter, indeed, the local strong solution in the $L^2$ space constructed by himself [23] was extended globally in time under the condition that both $a \in H_0^1(\Omega)$ (with compatibility conditions) and the other data are small in a sense, however, his method gave no information about the large time behavior of the solution.

The purpose of the present paper is to provide the global existence theorem for a unique strong solution $u(t)$ of (1.1), which satisfies the flux condition $\phi(u(t)) = 0$ and some decay properties with definite rates that seem to be optimal, for instance,

$$\|u(t)\|_{L^\infty(\Omega)} + \|\nabla u(t)\|_{L^n(\Omega)} = o\left(t^{-1/2}\right),$$

as $t \to \infty$, when the initial velocity $a$ is small enough in $L^n_\sigma(\Omega), n \geq 3$. The space $L^n$ is now well known as a reasonable class of initial data, from the viewpoint of scaling invariance, to find a global strong solution within the framework of $L^q$ theory. We derive further sharp decay properties of the solution $u(t)$ under the additional assumption $a \in L^1(\Omega) \cap L^n_\sigma(\Omega)$; for instance, the decay rate given above is improved as $O(t^{-n/2})$. For the proof, as is well known, it is crucial to establish the $L^q$-$L^r$ estimates of the Stokes semigroup

$$\|e^{-tA} f\|_{L^r(\Omega)} \leq C t^{-\alpha} \|f\|_{L^q(\Omega)}, \tag{1.3}$$

$$\|\nabla e^{-tA} f\|_{L^r(\Omega)} \leq C t^{-\alpha-1/2} \|f\|_{L^q(\Omega)}, \tag{1.4}$$

for all $t > 0$ and $f \in L^q_\sigma(\Omega)$, where $\alpha = (n/q - n/r)/2 \geq 0$. Recently for $n \geq 3$ Abels [1] has proved some partial results: (1.3) for $1 < q \leq r < \infty$ and (1.4) for $1 < q \leq r < n$. However, because of the lack of (1.4) for the most important case $q = r = n$, his results are not satisfactory for the construction of the global strong solution possessing various time-asymptotic behaviors as long as one follows the straightforward method of Kato [39] (without using duality arguments in [46], [7], [48], [49] and [37]). In this paper we consider the case $n \geq 3$ and prove

$$(1.3) \text{ for } 1 \leq q \leq r \leq \infty \ (q \neq \infty, r \neq 1),$$

and

$$(1.4) \text{ for } 1 \leq q \leq r \leq n \ (r \neq 1) \text{ or } 1 \leq q < n < r < \infty;$$

here, when $q = 1$, $f$ should be taken from $L^1(\Omega) \cap L^s_\sigma(\Omega)$ for some $s \in (1, \infty)$. Estimate (1.4) is thus available, in other words, for $r = n$ if $q = n$, for $r \in [q, \infty)$ if $q \in (1, n)$, and for $r \in (1, \infty)$ if $q = 1$.

Up to now we have the same global existence result as above for the whole space (Kato [39]), the half space (Ukai [65], Kozono [41]), bounded domains (Giga and Miyakawa [32]) and exterior domains (Iwashita [38]) since the $L^q$-$L^r$ estimates (1.3) and (1.4) are well established for these four types of domains. Let us give a brief survey on the literature concerning the $L^q$-$L^r$ estimates. For the whole space the Stokes semigroup is essentially the same as the heat semigroup because the Laplace operator commutes with the Helmholtz projection. For the half space Ukai [65] explicitly wrote down a solution formula of the Stokes system and derived (1.3) and (1.4) for $n \geq 2$ and $1 < q \leq r < \infty$. See also Borchers and Miyakawa [4] for (1.3) with $1 \leq q < r \leq \infty$ and the following literature concerning marginal cases, that is, (1.3) for $q = r = \infty$ and (1.4) for $q = r = 1$ or $\infty$: Giga, Matsui and Y. Shimizu [31], Y. Shimizu [59], Desch, Hieber and Prüss [17] and Shibata and S. Shimizu [58]. For bounded domains (1.3) and (1.4) are deduced from the result of Giga [30] on a characterization of the domains of fractional powers of the Stokes operator. In this case, moreover, an exponential decay property of the semigroup for large $t$ is available. For exterior domains with $n \geq 3$, based on (1.3) for $q = r$ due to Borchers and Sohr [9], some partial results were given by Iwashita [38], Giga and Sohr [33] and Borchers and Miyakawa [5]; in particular, Iwashita proved (1.3) for $1 < q \leq r < \infty$ and (1.4) for $1 < q \leq r \leq n$, which made it possible to construct a global solution. Later on, due to the following authors, (1.3) for $n \geq 2$, $1 \leq q \leq r \leq \infty$ $(q \neq \infty, r \neq 1)$ and (1.4) for $n \geq 2$, $1 \leq q \leq r \leq n$ $(r \neq 1)$ were also derived: Chen [13] $(n = 3)$, Shibata [57] $(n = 3)$, Borchers and Varnhorn [11] ($n = 2$, (1.3) for $q = r$), Dan and Shibata [15], [16] $(n = 2)$, Dan, Kobayashi and Shibata [14] $(n = 2, 3)$ and Maremonti and Solonnikov [50] $(n \geq 2)$.

In the proof of the $L^q$-$L^r$ estimates, it seems to be heuristically reasonable to combine some local decay properties near the aperture with the $L^q$-$L^r$ estimates of the Stokes semigroup for the half space by means of a localization procedure since the aperture domain $\Omega$ is obtained from $H_+ \cup H_-$ by a perturbation within a compact region. Indeed, Abels [1] used this idea that had been well developed by Iwashita [38] and, later, Kobayashi and Shibata [40] in the case of exterior domains. We should however note that the boundary $\partial\Omega$ is noncompact; thus, a difficulty is to deduce the sharp local energy decay estimate

$$\|e^{-tA}f\|_{W^{1,q}(\Omega_R)} \leq Ct^{-n/2q}\|f\|_{L^q(\Omega)}, \quad t \geq 1, \tag{1.5}$$

for $f \in L^q_\sigma(\Omega), 1 < q < \infty$, where $\Omega_R = \{x \in \Omega; |x| < R\}$, but this is the essential part of our proof (Lemma 5.3). Estimate (1.5) improves the local energy decay given by Abels [1], in which a little slower rate $t^{-n/2q+\varepsilon}$ was shown. In [1], similarly to Iwashita [38], a resolvent expansion around the origin $\lambda = 0$ was derived in some weighted function spaces. To this end, Abels made use of the Ukai formula of the Stokes semigroup for the half space ([65]) and, in order to estimate the Riesz operator appearing in this formula, he had to introduce Muckenhoupt weights, which caused some restrictions although his analysis itself is of interest. On the other hand, Kobayashi and Shibata [40] refined the proof of Iwashita in some sense and obtained the $L^q$-$L^r$ estimates of the Oseen semigroup for the 3-dimensional exterior domain. As a particular case, the result of [40] includes the estimates of the Stokes semigroup as well. In this paper we employ in principle the strategy developed by [40] (without using any weighted function space) and extend the method to general $n \geq 3$.

This paper consists of six sections. In the next section, after notation is fixed, we present the precise statement of our main results: Theorem 2.1 on the $L^q$-$L^r$ estimates of the Stokes semigroup, Theorem 2.2 on the global existence and decay properties of the Navier-Stokes flow, and Theorem 2.3 on some further asymptotic behaviors of the obtained flow under an additional summability assumption on initial data. We obtain an information about a pressure drop as well in the last theorem.

Section 3 is devoted to the investigation of the Stokes resolvent for the half space $H = H_+$ or $H_-$. We derive some regularity estimates near the origin $\lambda = 0$ of $(\lambda + A_H)^{-1} P_H f$ when $f \in L^q(H)$ has a bounded support, where $A_H = -P_H \Delta$ is the Stokes operator for the half space $H$ (for the notation, see Section 2). Although the obtained estimates do not seem to be optimal compared with those shown by [40] for the whole space, the results are sufficient for our aim and the proof is rather elementary: in fact, we represent the resolvent $(\lambda + A_H)^{-1}$ in terms of the semigroup $e^{-tA_H}$ and, with the aid of local energy decay properties of this semigroup, we have only to perform several integrations by parts and to estimate the resulting formulae. One needs neither Fourier analysis nor resolvent expansion.

In Section 4, based on the results for the half space, we proceed to the analysis of the Stokes resolvent for the aperture domain $\Omega$. To do so, in an analogous way to [38], [40] and [1], we first construct the resolvent $(\lambda + A)^{-1} Pf$ near the origin $\lambda = 0$ for $f \in L^q(\Omega)$ with bounded support by use of the operator $(\lambda + A_H)^{-1} P_H$, the Stokes flow in a bounded domain and a cut-off function together with the result of Bogovskiĭ [2] on the boundary value problem for the equation of continuity. And then, for the same $f$ as above, we deduce essentially the same regularity estimates near the origin $\lambda = 0$ of $(\lambda + A)^{-1} Pf$ as shown in Section 3.

In Section 5 we prove (1.5) and thereby (1.4) for $q = r \in (1, n]$ as well as (1.3) for $r = \infty$, from which the other cases follow. Some of the estimates obtained in Section 4 enable us to justify a representation formula of the semigroup $e^{-tA} Pf$ in $W^{1,q}(\Omega_R)$ in terms of the Fourier inverse transform of $\partial_s^m (is + A)^{-1} Pf$ when $f \in L^q(\Omega)$ has a bounded support, where $n = 2m + 1$ or $n = 2m + 2$ (see (5.3); we note that the formula is not valid for $n = 2$). We then appeal to the lemma due to Shibata ([56]; see also [40] and a recent development [58]), which tells us a relation between the regularity of a function at the origin and the decay property of its Fourier inverse image, so that we obtain another local energy decay estimate

$$\|e^{-tA} Pf\|_{W^{1,q}(\Omega_R)} \leq C t^{-n/2+\varepsilon} \|f\|_{L^q(\Omega)}, \quad t \geq 1, \tag{1.6}$$

for $f \in L^q(\Omega), 1 < q < \infty$, with bounded support, where $\varepsilon > 0$ is arbitrary (Lemma 5.1). Estimate (1.6) was shown in [1] only for solenoidal data $f \in L^q_\sigma(\Omega)$ with bounded support, from which (1.5) with the rate replaced by $t^{-n/2q+\varepsilon}$ follows through an interpolation argument. But it is crucial for the proof of (1.5) to use (1.6) even for data which are not solenoidal (so that the support of $Pf$ is unbounded). In order to deduce (1.5) from (1.6), we develop the method in [38] and [40] based on a localization argument using a cut-off function. In fact, we regard the Stokes flow for the aperture domain $\Omega$ as the sum of the Stokes flows for the half spaces $H_\pm$ and a certain perturbed flow. Since the Stokes flow for the half space enjoys the $L^q$-$L^\infty$ decay estimate with the rate $t^{-n/2q}$ (Borchers and Miyakawa [4]), our main task is to show (1.5) for the perturbation part. In contrast to the case of exterior domains, the support of the derivative of the cut-off function touches the boundary $\partial\Omega$; indeed, this difficulty occurs in all stages of localization procedures in the course of the proof (Sections 4 and 5) and thus we have to carry out such procedures carefully. Furthermore, the remainder term arising from the above-mentioned localization argument involves the pressure of the nonstationary Stokes system in the half space and, therefore, does not belong to any solenoidal function space. Hence, in order to treat this term, (1.6) is necessary for non-solenoidal data, while that is not the case for the exterior problem.

Once Theorem 2.1 is established, one can prove the existence part of Theorem 2.2 along the lines of Kato [39] (see also [26] and [32]) and therefore the proof may be omitted. Thus, in the final section, we derive various decay properties of the global strong solution as $t \to \infty$ to prove the remaining part of Theorem 2.2 and Theorem 2.3. This will be done by applying effectively the $L^q$-$L^r$ estimates. Recently Wiegner [66] has discussed in detail sharp decay properties of exterior Navier–Stokes flows. Our proof is somewhat different from his and seems to be elementary. When $a \in L^1(\Omega) \cap L^n_\sigma(\Omega)$, some decay rates are better than those shown by [66] since, unlike exterior Stokes flows, (1.4) is available for $1 \le q < n < r < \infty$.

Finally, we compare the result on $\nabla e^{-tA}$ with that for exterior Stokes flows from the viewpoint of coercive estimates of derivatives. For the proof of (1.4) there is another approach based on fractional powers of the Stokes operator. When $\Omega$ is an exterior domain ($n \ge 3$), Borchers and Miyakawa [5] developed such an approach and succeeded in the proof of

$$\|\nabla u\|_{L^q(\Omega)} \le C\|A^{1/2}u\|_{L^q(\Omega)}, \quad u \in D(A_q^{1/2}), \tag{1.7}$$

for $1 < q < n$ (this restriction is optimal as pointed out by themselves [6]), which implies (1.4) for $q \le r < n$. Independently, as mentioned, Iwashita [38] derived (1.4) for $q \le r \le n$ and, later, Maremonti and Solonnikov [50] showed that the restriction $r \le n$ cannot be improved for exterior domains. In our case of aperture domains, we have (1.4) for $q < n < r < \infty$, which is a consequence of the estimate due to Farwig and Sohr ([22, Theorem 2.5])

$$\|\nabla^2 u\|_{L^q(\Omega)} \le C\|Au\|_{L^q(\Omega)}, \quad u \in D(A_q), \tag{1.8}$$

for $1 < q < n$ together with an embedding property ([22, Lemma 3.1]); we mention that (1.8) holds true for $n = 2$ as well. This argument does not work for the exterior problem because (1.8) is valid only for $1 < q < n/2$ ($n \ge 3$) as shown by Borchers and Sohr [9] (the restriction on $q$ is again optimal by, for instance, [6]). Thus, as for (1.8), we have the better result. We wish we could expect (1.7) for every $q$, which would imply (1.4) for $1 < q \le r < \infty$; however, so far, no attempts have been made at the boundedness of purely imaginary powers of the Stokes operator (see [30] and [33] for bounded and exterior domains) and, unless $q = 2$, estimate (1.7) remains open.

## 2. Results

Before stating our main results, we introduce notation used throughout this paper. We denote upper and lower half spaces by $H_\pm = \{x \in \boldsymbol{R}^n; \pm x_n > 1\}$, and sometimes write $H = H_+$ or $H_-$ to state some assertions for the half space. Set $B_R = \{x \in \boldsymbol{R}^n; |x| < R\}$ for $R > 0$. Let $\Omega \subset \boldsymbol{R}^n$ be a given aperture domain with smooth boundary $\partial\Omega$, namely, there is $R_0 > 1$ so that

$$\Omega \setminus B_{R_0} = (H_+ \cup H_-) \setminus B_{R_0};$$

in what follows we fix such $R_0$. Since $\Omega$ should be connected, there are some apertures and one can take two disjoint subdomains $\Omega_\pm$ and a smooth $(n-1)$-dimensional manifold $M$ such that $\Omega = \Omega_+ \cup M \cup \Omega_-, \Omega_\pm \setminus B_{R_0} = H_\pm \setminus B_{R_0}$ and $M \cup \partial M = \partial\Omega_+ \cap \partial\Omega_- \subset \overline{B_{R_0}}$. We set $\Omega_R = \Omega \cap B_R$ and $H_R = H \cap B_R$, which is one of $H_{\pm,R} = H_\pm \cap B_R$, for $R > 1$.

For a domain $G \subset \boldsymbol{R}^n$, integer $j \geq 0$ and $1 \leq q \leq \infty$, we denote by $W^{j,q}(G)$ the standard $L^q$-Sobolev space with norm $\|\cdot\|_{j,q,G}$ so that $L^q(G) = W^{0,q}(G)$ with norm $\|\cdot\|_{q,G}$. The space $W_0^{j,q}(G)$ is the completion of $C_0^\infty(G)$, the class of $C^\infty$ functions having compact support in $G$, in the norm $\|\cdot\|_{j,q,G}$, and $W^{-j,q}(G)$ stands for the dual space of $W_0^{j,q/(q-1)}(G)$ with norm $\|\cdot\|_{-j,q,G}$. For simplicity, we use the abbreviations $\|\cdot\|_q$ for $\|\cdot\|_{q,\Omega}$ and $\|\cdot\|_{j,q}$ for $\|\cdot\|_{j,q,\Omega}$ when $G = \Omega$. We often use the same symbols for denoting the vector and scalar function spaces if there is no confusion. It is convenient to introduce a Banach space

$$L^q_{[R]}(G) = \{u \in L^q(G); \text{supp } u \subset \overline{G_R}\}, \quad G = \Omega \text{ or } H,$$

for $R > 1$, where supp $u$ denotes the support of the function $u$. For a Banach space $X$ we denote by $B(X)$ the Banach space which consists of all bounded linear operators from $X$ into itself.

Given $R \geq R_0$, we take (and fix) two cut-off functions $\psi_{\pm,R}$ satisfying

$$\psi_{\pm,R} \in C^\infty(\boldsymbol{R}^n; [0,1]), \quad \psi_{\pm,R}(x) = \begin{cases} 1 & \text{in } H_\pm \setminus B_{R+1}, \\ 0 & \text{in } H_\mp \cup B_R. \end{cases} \tag{2.1}$$

In some localization procedures with use of the cut-off functions above, the bounded domain of the form

$$D_{\pm,R} = \{x \in H_\pm; R < |x| < R+1\}$$

appears, and for this we need the following result of Bogovskiĭ [2] which provides a certain solution having an optimal regularity of the boundary value problem for $\nabla \cdot u = f$ with $u = 0$ on the boundary (see also Pileckas [54, Lemma 1], Borchers and Sohr [10, Theorem 2.4 (a)(b)(c)] and Galdi [28, Chapter III]): there is a linear operator $S_{\pm,R}$ from $C_0^\infty(D_{\pm,R})$ to $C_0^\infty(D_{\pm,R})^n$ such that for $1 < q < \infty$ and integer $j \geq 0$

$$\|\nabla^{j+1} S_{\pm,R} f\|_{q,D_{\pm,R}} \leq C \|\nabla^j f\|_{q,D_{\pm,R}}, \tag{2.2}$$

with $C = C(R,q,j) > 0$ independent of $f \in C_0^\infty(D_{\pm,R})$ (where $\nabla^j$ denotes all the $j$-th derivatives); and

$$\nabla \cdot S_{\pm,R} f = f,$$

for all $f \in C_0^\infty(D_{\pm,R})$ with $\int_{D_{\pm,R}} f(x)dx = 0$. By (2.2) the operator $S_{\pm,R}$ extends uniquely to a bounded operator from $W_0^{j,q}(D_{\pm,R})$ to $W_0^{j+1,q}(D_{\pm,R})^n$.

For $G = \Omega, H$ and a smooth bounded domain ($n \geq 2$), let $C_{0,\sigma}^{\infty}(G)$ be the set of all solenoidal (divergence free) vector fields whose components belong to $C_0^{\infty}(G)$, and $L_{\sigma}^q(G)$ the completion of $C_{0,\sigma}^{\infty}(G)$ in the norm $\|\cdot\|_{q,G}$. If, in particular, $G = \Omega$, then the space $L_{\sigma}^q(\Omega)$ is characterized as (1.2). The space $L^q(G)$ of vector fields admits the Helmholtz decomposition

$$L^q(G) = L_{\sigma}^q(G) \oplus L_{\pi}^q(G), \quad 1 < q < \infty,$$

with $L_{\pi}^q(G) = \{\nabla p \in L^q(G); p \in L_{loc}^q(\overline{G})\}$; see [27], [60] for bounded domains, [4], [52] for $G = H$ and [22], [53] for $G = \Omega$. Let $P_{q,G}$ be the projection operator from $L^q(G)$ onto $L_{\sigma}^q(G)$ associated with the decomposition above. Then the Stokes operator $A_{q,G}$ is defined by the solenoidal part of the Laplace operator, that is,

$$D(A_{q,G}) = W^{2,q}(G) \cap W_0^{1,q}(G) \cap L_{\sigma}^q(G), \quad A_{q,G} = -P_{q,G}\Delta,$$

for $1 < q < \infty$. The dual operator $A_{q,G}^*$ of $A_{q,G}$ coincides with $A_{q/(q-1),G}$ on $L_{\sigma}^q(G)^* = L_{\sigma}^{q/(q-1)}(G)$. We use, for simplicity, the abbreviations $P_q$ for $P_{q,\Omega}$ and $A_q$ for $A_{q,\Omega}$, and the subscript $q$ is also often omitted if there is no confusion. The Stokes operator enjoys the parabolic resolvent estimate

$$\|(\lambda + A_G)^{-1}\|_{B(L_{\sigma}^q(G))} \leq C_{\varepsilon}/|\lambda|, \tag{2.3}$$

for $|\arg\lambda| \leq \pi - \varepsilon$ ($\lambda \neq 0$), where $\varepsilon > 0$ is arbitrarily small; see [29], [61] for bounded domains, [52], [4], [19], [20], [17] for $G = H$ and [22] for $G = \Omega$. Estimate (2.3) implies that the operator $-A_G$ generates a bounded analytic semigroup $\{e^{-tA_G}; t \geq 0\}$ of class $(C_0)$ in each $L_{\sigma}^q(G), 1 < q < \infty$. We write $E(t) = e^{-tA_H}$, which is one of $E_{\pm}(t) = e^{-tA_{H_{\pm}}}$.

The first theorem provides the $L^q$-$L^r$ estimates of the Stokes semigroup $e^{-tA}$ for the aperture domain $\Omega$.

**Theorem 2.1.** *Let $n \geq 3$.*

1. *Let $1 \leq q \leq r \leq \infty$ ($q \neq \infty, r \neq 1$). There is a constant $C = C(\Omega, n, q, r) > 0$ such that (1.3) holds for all $t > 0$ and $f \in L_{\sigma}^q(\Omega)$ unless $q = 1$; when $q = 1$, the assertion remains true if $f$ is taken from $L^1(\Omega) \cap L_{\sigma}^s(\Omega)$ for some $s \in (1, \infty)$.*
2. *Let $1 \leq q \leq r \leq n$ ($r \neq 1$) or $1 \leq q < n < r < \infty$. There is a constant $C = C(\Omega, n, q, r) > 0$ such that (1.4) holds for all $t > 0$ and $f \in L_{\sigma}^q(\Omega)$ unless $q = 1$; when $q = 1$, the assertion remains true if $f$ is taken from $L^1(\Omega) \cap L_{\sigma}^s(\Omega)$ for some $s \in (1, \infty)$.*
3. *Let $1 < q < \infty$ and $f \in L_{\sigma}^q(\Omega)$. Then*

$$\|e^{-tA}f\|_r = o(t^{-\alpha}) \quad \begin{cases} \text{as } t \to 0 & \text{if } q < r \leq \infty, \\ \text{as } t \to \infty & \text{if } q \leq r \leq \infty, \end{cases} \tag{2.4}$$

$$\|\nabla e^{-tA}f\|_r = o(t^{-\alpha-1/2}) \quad \begin{cases} \text{as } t \to 0 & \text{if } q \leq r \leq \infty, \\ \text{as } t \to \infty & \text{if } q \leq r \leq n, \; q < n < r < \infty, \end{cases} \tag{2.5}$$

*where $\alpha = (n/q - n/r)/2$. Furthermore, for each precompact set $K$ in $L_{\sigma}^q(\Omega)$ every convergence above is uniform with respect to $f \in K$.*

**Remark 2.1.** Estimate (1.4) for large $t$ is not proved in the following cases: (i) $n < q = r < \infty$, (ii) $n \le q < r < \infty$. For the case (i), the decay rate $t^{-n/2q}$ will be shown in Lemma 5.4. Since we have (1.4) for $q < n < r < \infty$, a better decay rate than $t^{-n/2q}$ can be derived for the case (ii) through an interpolation argument; however, we do not know optimal decay rates of $\nabla e^{-tA}$ in the cases (i) and (ii). According to Maremonti and Solonnikov [50], the decay rate $t^{-n/2q}$ is optimal for exterior Stokes flows whenever $r > n$.

**Remark 2.2.** Let $1 \le q \le r \le \infty$ $(q \ne \infty, r \ne 1)$. The $L^q$-$L^r$ estimate for $\partial_t e^{-tA}$ with the rate $t^{-\alpha-1}$ is nothing but a simple corollary to (1.3). In fact, for example,

$$\|\partial_t e^{-tA} f\|_\infty \le C t^{-n/2s} \|A e^{-(t/2)A} f\|_s \le C t^{-n/2-1} \|f\|_1,$$

for $t > 0$ and $f \in L^1(\Omega) \cap L^s_\sigma(\Omega)$.

By use of the Stokes operator $A$, one can formulate the problem (1.1) subject to the vanishing flux condition

$$\phi(u(t)) = \int_M N \cdot u(t) d\sigma = 0, \quad t \ge 0, \tag{2.6}$$

as the Cauchy problem

$$\partial_t u + Au + P(u \cdot \nabla u) = 0, \quad t > 0; u(0) = a, \tag{2.7}$$

in $L^q_\sigma(\Omega)$. Given $a \in L^n_\sigma(\Omega)$ and $0 < T \le \infty$, a measurable function $u$ defined on $\Omega \times (0, T)$ is called a strong solution of (1.1) with (2.6) on $(0, T)$ if $u$ is of class

$$u \in C([0, T); L^n_\sigma(\Omega)) \cap C(0, T; D(A_n)) \cap C^1(0, T; L^n_\sigma(\Omega))$$

together with $\lim_{t \to 0} \|u(t) - a\|_n = 0$ and satisfies (2.7) for $0 < t < T$ in $L^n_\sigma(\Omega)$.

The next theorem tells us the global existence of a strong solution with several decay properties provided that $\|a\|_n$ is small enough.

**Theorem 2.2.** *Let $n \ge 3$. There is a constant $\delta = \delta(\Omega, n) > 0$ with the following property: if $a \in L^n_\sigma(\Omega)$ satisfies $\|a\|_n \le \delta$, then the problem* (1.1) *with* (2.6) *admits a unique strong solution $u(t)$ on $(0, \infty)$, which enjoys*

$$\|u(t)\|_r = o\left(t^{-1/2+n/2r}\right) \quad \textit{for } n \le r \le \infty, \tag{2.8}$$

$$\|\nabla u(t)\|_n = o\left(t^{-1/2}\right), \tag{2.9}$$

$$\|\partial_t u(t)\|_n + \|Au(t)\|_n = o\left(t^{-1}\right), \tag{2.10}$$

*as $t \to \infty$.*

**Remark 2.3.** When one prescribes a nontrivial flux

$$\phi(u(t)) = F(t) \in C^{1,\theta}([0, T]),$$

with some $\theta > 0$ and $T > 0$, there is $T_* \in (0, T]$ such that the problem (1.1) with the flux condition admits a unique strong solution on $(0, T_*)$ provided that $a \in L^n(\Omega)$ satisfies the compatibility conditions $\nabla \cdot a = 0, \nu \cdot a|_{\partial\Omega} = 0$ and $\phi(a) = F(0)$. This improves a related result of Franzke [25] and can be proved in the same manner as the proof of Theorem 2.2 with the aid of the auxiliary function of Heywood [34, Lemma 11], which is used for the reduction of the problem to an equivalent one with the vanishing flux condition (2.6). As is well known, (2.4) and (2.5) as $t \to 0$ play an important role for the construction of the above local solution.

**Remark 2.4.** The solution obtained in Theorem 2.2 is unique within the class

$$u \in C([0,\infty); L^n_\sigma(\Omega)), \quad \nabla u \in C(0,\infty; L^n(\Omega)),$$

without assuming any behavior (6.4) near $t = 0$ as pointed out by Brezis [12]. For the proof, one needs the final assertion of Theorem 2.1 on the uniform behavior of the semigroup as $t \to 0$ on each precompact set $K$ in $L^n_\sigma(\Omega)$ together with the theory of local strong solutions mentioned in the previous remark (with $\phi(u) = F = 0$). In fact, it follows from the above property of the semigroup that the length of the existence interval of the local solution can be taken uniformly with respect to $a \in K$ and that the convergence (6.4) of the local solution as $t \to 0$ is also uniform with respect to $a \in K$. These two facts combined with the classical uniqueness theorem of Fujita–Kato type [26] (assuming some behaviors in (6.4) near $t = 0$) imply the desired uniqueness result.

**Remark 2.5.** Consider the 3-dimensional stationary Navier–Stokes problem

$$w \cdot \nabla w = \Delta w - \nabla \pi, \quad \nabla \cdot w = 0,$$

in $\Omega$ subject to $w|_{\partial\Omega} = 0$ and a nontrivial flux condition $\phi(w) = \gamma \in \boldsymbol{R}$. When $|\gamma|$ is small enough, there is a unique solution $w$ such that $w \in L^q(\Omega)$ for $3/2 < q \leq 6$ and $\nabla w \in L^r(\Omega)$ for $1 < r \leq 2$ with $\|\nabla w\|_2^2 = \gamma[\pi]$; see Galdi [28]. By use of Theorem 2.1 it is possible to show the asymptotic stability of the small stationary solution $w$ of the class above for small initial disturbance in $L^3_\sigma(\Omega)$ in the sense that the disturbance $u(t)$ decays like (2.8) and (2.9) as $t \to \infty$. In fact, the above summability properties of $\nabla w$ allow us to deal with the term $P(w \cdot \nabla u + u \cdot \nabla w)$ as a simple perturbation of the Stokes operator, as was done by Chen [13, Lemma 3.1] and Borchers and Miyakawa [7, Theorem 3.13]; see Remark 6.1.

The final theorem shows further decay properties of the global solution when we additionally impose $L^1$-summability on the initial data.

**Theorem 2.3.** *Let $n \geq 3$. There is a constant $\eta = \eta(\Omega, n) \in (0, \delta]$ with the following property: if $a \in L^1(\Omega) \cap L^n_\sigma(\Omega)$ satisfies $\|a\|_n \leq \eta$, then the solution $u(t)$ obtained in Theorem 2.2 and the associated pressure $p(t)$ enjoy*

$$\|u(t)\|_r = O\left(t^{-(n-n/r)/2}\right) \quad \text{for } 1 < r \leq \infty, \tag{2.11}$$

$$\|\nabla u(t)\|_r = O\left(t^{-(n-n/r)/2-1/2}\right) \quad \text{for } 1 < r < \infty, \tag{2.12}$$

$$\|\partial_t u(t)\|_r + \|Au(t)\|_r = O\left(t^{-(n-n/r)/2-1}\right) \quad \text{for } 1 < r < \infty, \tag{2.13}$$

$$\|\nabla^2 u(t)\|_r + \|\nabla p(t)\|_r = O\left(t^{-(n-n/r)/2-1}\right) \quad \text{for } 1 < r < n, \tag{2.14}$$

*as $t \to \infty$. Moreover, for each $t > 0$ there exist two constants $p_\pm(t) \in \boldsymbol{R}$ such that $p(t) - p_\pm(t) \in L^r(\Omega_\pm)$ with*

$$\|p(t) - p_\pm(t)\|_{r,\Omega_\pm} = O\left(t^{-(n-n/r)/2-1/2}\right) \text{ for } n/(n-1) < r < \infty, \tag{2.15}$$

$$|p_+(t) - p_-(t)| = O\left(t^{-n/2-1/2+\varepsilon}\right), \tag{2.16}$$

*as $t \to \infty$, where $\varepsilon > 0$ is arbitrarily small.*

**Remark 2.6.** Indeed $\nabla u(t) \in L^r(\Omega)$ for $r > n$ even in Theorem 2.2, but we have asserted nothing about their decay rates since they do not seem to be optimal; see Remark 2.1 for the Stokes flow. On the other hand, in Theorem 2.3 the decay rates of $\nabla u(t)$ in $L^r(\Omega)$ for $r > n$ are better than $t^{-n/2}$ for exterior Navier–Stokes flows shown by Wiegner [66]. Taking Theorem 5.1 of [17] for the Stokes flow in the half space into account, we would not expect $u(t) \in L^1(\Omega)$ in general. Thus the decay rates obtained in Theorem 2.3 seem to be optimal; that is, for example, $\|u(t)\|_\infty = o(t^{-n/2})$ would not hold true. Concerning the exterior problem, Kozono [42], [43] made it clear that the Stokes and/or Navier–Stokes flows possess $L^1$-summability and more rapid decay properties than (2.11) only in a special situation.

**Remark 2.7.** In Theorem 2.2 one could not define a pressure drop (see Farwig [18, Remark 2.2]) since the solution does not always belong to $L^r(\Omega)$ for $r < n$. Due to the additional summability assumption on the initial data, we obtain in Theorem 2.3 the pressure drop written in the form

$$[p(t)] = p_+(t) - p_-(t) = \int_\Omega (\partial_t u + u \cdot \nabla u - u)(t) \cdot w dx,$$

where $w \in W^{2,q}(\Omega)$, $n/(n-1) < q < \infty$, is a unique solution (given by [22]) of the auxiliary problem

$$w - \Delta w + \nabla \pi = 0, \quad \nabla \cdot w = 0,$$

in $\Omega$ subject to $w|_{\partial\Omega} = 0$ and $\phi(w) = 1$. In fact, the formula above is derived from the relations

$$\int_\Omega w \cdot \nabla p(t) dx = -[p(t)]\phi(w) = -[p(t)],$$

$$\int_\Omega u(t) \cdot \nabla \pi dx = -[\pi]\phi(u(t)) = 0.$$

## 3. The Stokes resolvent for the half space

The resolvent $v = (\lambda + A_H)^{-1} P_H f$ together with the associated pressure $\pi$ solves the system

$$\lambda v - \Delta v + \nabla \pi = f, \quad \nabla \cdot v = 0,$$

in the half space $H = H_+$ or $H_-$ subject to $v|_{\partial H} = 0$ for the external force $f \in L^q(H), 1 < q < \infty$, and $\lambda \in \boldsymbol{C} \setminus (-\infty, 0]$. In this section we are concerned with the analysis of $v$ near $\lambda = 0$. Our method is quite different from Abels [1]. One needs the following local energy decay estimate of the semigroup $E(t) = e^{-tA_H}$, which is a simple consequence of (1.3) for $\Omega = H$.

**Lemma 3.1.** *Let $n \geq 2, 1 < q < \infty, d > 1$ and $R > 1$. For any small $\varepsilon > 0$ and integer $k \geq 0$ there is a constant $C = C(n, q, d, R, \varepsilon, k) > 0$ such that*

$$\|\nabla^j \partial_t^k E(t) P_H f\|_{q,H_R} \leq C t^{-j/2-k}(1+t)^{-n/2+\varepsilon}\|f\|_{q,H}, \tag{3.1}$$

*for $t > 0, f \in L^q_{[d]}(H)$ and $j = 0, 1, 2$.*

*Proof.* We make use of the estimate

$$\|\nabla^j u\|_{r,H} \leq C\|A_H^{j/2} u\|_{r,H}, \quad u \in D(A_{r,H}^{j/2}), \tag{3.2}$$

for $1 < r < \infty$ and $j = 1, 2$ (Borchers and Miyakawa [4]). For $1 < p \leq q \leq r < \infty$ it follows from (1.3) for $\Omega = H$, (3.2) and a property of the analytic semigroup that

$$\begin{aligned} \|\nabla^j \partial_t^k E(t) P_H f\|_{q,H_R} &\leq C\|A_H^{j/2+k} E(t) P_H f\|_{r,H} \\ &\leq C t^{-j/2-k}\|E(t/2) P_H f\|_{r,H} \\ &\leq C t^{-j/2-k-(n/p-n/r)/2}\|f\|_{p,H} \\ &\leq C t^{-j/2-k-(n/p-n/r)/2}\|f\|_{q,H}, \end{aligned}$$

for $t > 0, f \in L^q_{[d]}(H)$ and $j = 0, 1, 2$. This estimate with $p = q = r$ implies (3.1) for $0 < t < 1$. We may assume that $0 < \varepsilon/n < \min\{1/q, 1 - 1/q\}$; and then one can take $p$ and $r$ so that $1 - 1/p = 1/r = \varepsilon/n$ and $p < q < r$. Then the estimate above yields (3.1) for $t \geq 1$. This completes the proof. □

Lemma 3.1 is sufficient for our analysis of the resolvent in this section, but the local energy decay estimate of the following form will be used in section 5.

**Lemma 3.2.** *Let $n \geq 2, 1 < q < \infty$ and $R > 1$. Then there is a constant $C = C(n, q, R) > 0$ such that*

$$\|E(t) f\|_{2,q,H_R} + \|\partial_t E(t) f\|_{q,H_R} \leq C(1+t)^{-n/2q}\|f\|_{D(A_{q,H})}, \tag{3.3}$$

*for $t \geq 0$ and $f \in D(A_{q,H})$.*

*Proof.* The left hand side of (3.3) is bounded from above by

$$C(\|A_H E(t) f\|_{q,H} + \|E(t) f\|_{q,H}) \leq C\|f\|_{D(A_{q,H})},$$

which implies (3.3) for $0 \leq t < 1$. For $t \geq 1$ it follows from (1.3) for $\Omega = H$ with $r = \infty$ that

$$\|E(t) f\|_{q,H_R} \leq C\|E(t) f\|_{\infty,H} \leq C t^{-n/2q}\|f\|_{q,H}.$$

The other terms

$$\|\nabla^j E(t) f\|_{q,H_R} \leq C\|A_H^{j/2} E(t) f\|_{r,H} \leq C t^{-j/2}\|E(t/2) f\|_{r,H} \quad (j = 1, 2),$$

$$\|\partial_t E(t) f\|_{q,H_R} \leq C t^{-1}\|E(t/2) f\|_{r,H},$$

decay more rapidly since we can take $r \in (q, \infty)$ above as large as we want. The proof is complete. □

We next employ Lemma 3.1 to show some regularity estimates near $\lambda = 0$ of the Stokes resolvent in the localized space $W^{2,q}(H_R)$.

**Lemma 3.3.** *Let $n \geq 3, 1 < q < \infty, d > 1$ and $R > 1$. Given $f \in L^q_{[d]}(H)$, set $v(\lambda) = (\lambda + A_H)^{-1} P_H f$. For any small $\varepsilon > 0$ there is a constant $C = C(n, q, d, R, \varepsilon) > 0$ such that*

$$|\lambda|^\beta \|\partial_\lambda^m v(\lambda)\|_{2,q,H_R} + \sum_{k=0}^{m-1} \|\partial_\lambda^k v(\lambda)\|_{2,q,H_R} \leq C\|f\|_{q,H}, \tag{3.4}$$

*for Re $\lambda \geq 0$ ($\lambda \neq 0$) and $f \in L^q_{[d]}(H)$, where*

$$m = \begin{cases} (n-1)/2 & \text{if } n \text{ is odd,} \\ n/2 - 1 & \text{if } n \text{ is even,} \end{cases}$$

$$\beta = \beta(\varepsilon) = 1 + m - \frac{n}{2} + \varepsilon = \begin{cases} 1/2 + \varepsilon & \text{if } n \text{ is odd,} \\ \varepsilon & \text{if } n \text{ is even.} \end{cases}$$

*Furthermore, we have*

$$\sup \left\{ \frac{\|v(\lambda) - w\|_{2,q,H_R}}{\|f\|_{q,H}};\ f \neq 0, f \in L^q_{[d]}(H) \right\} \to 0, \tag{3.5}$$

*as $\lambda \to 0$ with Re $\lambda \geq 0$, where*

$$w = \int_0^\infty E(t) P_H f\,dt.$$

*Proof.* We recall the formula

$$v(\lambda) = (\lambda + A_H)^{-1} P_H f = \int_0^\infty e^{-\lambda t} E(t) P_H f\,dt, \tag{3.6}$$

which is valid in $L^q_\sigma(H)$ for Re $\lambda > 0$ and $f \in L^q(H)$. In the other region $\{\lambda \in \boldsymbol{C} \setminus (-\infty, 0]; \text{Re } \lambda \leq 0\}$ we usually utilize the analytic extension of the semigroup $\{E(t); \text{Re } t > 0\}$ to obtain the similar formula. For the case Re $\lambda = 0$ ($\lambda \neq 0$) which is important for us, however, thanks to the local energy decay property (3.1), the formula (3.6) remains valid in the localized space $L^q(H_R)$ for $f \in L^q_{[d]}(H)$ (the function $w$ in (3.5) is well-defined in $L^q(H_R)$ by the same reasoning). We thus obtain from (3.1)

$$\|\nabla^j \partial_\lambda^k v(\lambda)\|_{q,H_R} \leq \int_0^\infty t^k \|\nabla^j E(t) P_H f\|_{q,H_R} dt \leq C\|f\|_{q,H},$$

provided that

$$j = 0, 1 \quad \text{if } k = 0; \quad j = 0, 1, 2 \quad \text{if } n \geq 5,\ 1 \leq k \leq m - 1;$$
$$j = 2 \quad \text{if } k = m, n = 2m + 1; \quad j = 1, 2 \quad \text{if } k = m, n = 2m + 2.$$

For $\{k, j\} = \{0, 2\}$ we have only to use (3.2) together with (2.3) to see that

$$\|\nabla^2 v(\lambda)\|_{q,H_R} \leq C\|A_H(\lambda + A_H)^{-1} P_H f\|_{q,H} \leq C\|f\|_{q,H}.$$

The remaining case $k = m$ is the most important part of (3.4). Since

$$\begin{aligned}\|\partial_\lambda^m v(\lambda)\|_{2,q,H_R} &\le Cm!\|(\lambda + A_H)^{-(m+1)} P_H f\|_{D(A_{q,H})}\\ &\le Cm!\{|\lambda|^{-m} + |\lambda|^{-(m+1)}\}\|f\|_{q,H},\end{aligned}$$

we have the assertion for $|\lambda| \ge 1$. For $0 < |\lambda| < 1$ and odd $n$ (resp. even $n$), we have already shown the estimate as above when $j = 2$ (resp. $j = 1, 2$). Thus, let $j = 0$ or $1$ for $n = 2m + 1$ and $j = 0$ for $n = 2m + 2$. We divide the integral of (3.6) into two parts

$$\partial_\lambda^m v(\lambda) = \left\{\int_0^{1/|\lambda|} + \int_{1/|\lambda|}^\infty\right\} e^{-\lambda t}(-t)^m E(t) P_H f\,dt = w_1(\lambda) + w_2(\lambda).$$

Then (3.1) implies

$$\|\nabla^j w_1(\lambda)\|_{q,H_R} \le C|\lambda|^{-\beta + j/2}\|f\|_{q,H},$$

for $f \in L^q_{[d]}(H)$. On the other hand, by integration by parts we get

$$w_2(\lambda) = \frac{e^{-\lambda/|\lambda|}}{\lambda}\left(\frac{-1}{|\lambda|}\right)^m E\left(\frac{1}{|\lambda|}\right) P_H f + \int_{1/|\lambda|}^\infty \frac{e^{-\lambda t}}{\lambda}\partial_t\left[(-t)^m E(t) P_H f\right] dt,$$

in $L^q(H_R)$ since (3.1) implies $\lim_{t\to\infty} t^m \|E(t) P_H f\|_{q,H_R} = 0$. With the aid of (3.1) again we see that

$$\begin{aligned}&\|\nabla^j w_2(\lambda)\|_{q,H_R}\\ \le\ &\frac{1}{|\lambda|^{m+1}}\|\nabla^j E(1/|\lambda|) P_H f\|_{q,H_R} + \frac{1}{|\lambda|}\int_{1/|\lambda|}^\infty \|\nabla^j \partial_t\left[t^m E(t) P_H f\right]\|_{q,H_R} dt\\ \le\ &C|\lambda|^{-\beta + j/2}\|f\|_{q,H},\end{aligned}$$

for $f \in L^q_{[d]}(H)$. Collecting the estimates above leads us to (3.4). We next show (3.5). Since $|e^{-\lambda t} - 1| \le 2^{1-\theta}|\lambda|^\theta t^\theta$ for $\mathrm{Re}\,\lambda \ge 0$ and $\theta \in (0, 1]$, we have

$$\|\nabla^j(v(\lambda) - w)\|_{q,H_R} \le 2^{1-\theta}|\lambda|^\theta \int_0^\infty t^\theta \|\nabla^j E(t) P_H f\|_{q,H_R} dt,$$

for $j = 0, 1, 2$. From (3.1) together with a suitable choice of $\theta$ (for instance, $\theta < 1/2$ for $n = 3$), we conclude (3.5). □

**Remark 3.1.** When $n = 2$, one can show $|\lambda|^\beta \|v\|_{2,q,H_R} \le C\|f\|_{q,H}$ (with $\beta = \varepsilon$) which corresponds to (3.4) with $m = 0$. However, this will not help us since our key formula (5.3) is not valid for $m = 0$.

**Remark 3.2.** The Green tensor associated with the Stokes semigroup $E(t)$ for the half space (as well as the projection $P_H$) was explicitly given by Solonnikov [61], Maremonti and Solonnikov [50, section 2]. In view of the simple relation $\int_0^\infty (4\pi t)^{-n/2} e^{-|x|^2/4t} dt = \Gamma(n/2)/2(n-2)\pi^{n/2}|x|^{n-2}$ for $n \ge 3$, the function $w$ in (3.5) is the solution written by the Green tensor for the stationary Stokes problem in $H$ and, thereby, we know the class of $w$ (for the latter Green tensor, see for instance [28]).

Finally, we derive further information on the regularity of the resolvent along the imaginary axis.

**Lemma 3.4.** *Let $n \geq 3, 1 < q < \infty, d > 1$ and $R > 1$. Set*

$$\Phi_H^{(k)}(s) = \partial_s^k (is + A_H)^{-1} P_H \quad (s \in \boldsymbol{R} \setminus \{0\},\ k = m \text{ or } m-1),$$

*where $i = \sqrt{-1}$. Then, for any small $\varepsilon > 0$, there is a constant $C = C(n, q, d, R, \varepsilon) > 0$ such that*

$$\|\Phi_H^{(m)}(s+h)f - \Phi_H^{(m)}(s)f\|_{2,q,H_R} \leq C|h||s|^{-\beta-1}\|f\|_{q,H}, \tag{3.7}$$

$$\|\Phi_H^{(m-1)}(s+h)f - \Phi_H^{(m-1)}(s)f\|_{2,q,H_R} \leq C|h||s|^{-\beta}\|f\|_{q,H}, \tag{3.8}$$

*for $h \in \boldsymbol{R}, |s| > 2|h|$ and $f \in L^q_{[d]}(H)$, where $m$ and $\beta = \beta(\varepsilon)$ are the same as in Lemma 3.3.*

*Proof.* Estimate (3.8) is a direct consequence of (3.4). In fact, we see that

$$\begin{aligned}\|\Phi_H^{(m-1)}(s+h)f - \Phi_H^{(m-1)}(s)f\|_{2,q,H_R} &\leq \left| \int_s^{s+h} \|\Phi_H^{(m)}(\tau)f\|_{2,q,H_R} d\tau \right| \\ &\leq C\|f\|_{q,H} \left| \int_s^{s+h} |\tau|^{-\beta} d\tau \right|,\end{aligned}$$

which together with the relation $|s+h| \geq |s| - |h| \geq |s|/2$ implies (3.8). We next show (3.7). By (3.6) with Re $\lambda = 0$ in $L^q(H_R)$ we have

$$\begin{aligned}&\Phi_H^{(m)}(s+h)f - \Phi_H^{(m)}(s)f \\ &= (-i)^m \left\{ \int_0^{1/|s|} + \int_{1/|s|}^\infty \right\} e^{-ist}(e^{-iht} - 1)t^m E(t)P_H f dt = (-i)^m(w_1 + w_2).\end{aligned}$$

For the convenience we introduce the function

$$F_k(t) = \partial_t^k [t^m E(t) P_H f], \quad k \geq 0.$$

We then deduce from (3.1)

$$\|F_k(t)\|_{2,q,H_R} \leq Ct^{-k+m-1}(1+t)^{-n/2+1+\varepsilon}\|f\|_{q,H}, \tag{3.9}$$

for $t > 0$ and $f \in L^q_{[d]}(H)$. Taking $|e^{-iht} - 1| \leq |h|t$ into account, we see from (3.9) that

$$\begin{aligned}\|w_1\|_{2,q,H_R} &\leq |h| \int_0^{1/|s|} t\|F_0(t)\|_{2,q,H_R} dt \\ &\leq C|h|\|f\|_{q,H} \int_0^{1/|s|} t^{1+m-n/2+\varepsilon} dt \\ &\leq C|h||s|^{-\beta-1}\|f\|_{q,H},\end{aligned}$$

for $f \in L^q_{[d]}(H)$. By integration by parts we split $w_2 = w_{21} + w_{22} + w_{23}$, where

$$w_{21} = \frac{ih}{s(s+h)} e^{-i(s+h)/|s|} F_0\left(\frac{1}{|s|}\right) - \frac{i}{s} e^{-is/|s|}(e^{-ih/|s|} - 1)F_0\left(\frac{1}{|s|}\right),$$

$$w_{22} = \frac{ih}{s(s+h)} \int_{1/|s|}^{\infty} e^{-i(s+h)t} F_1(t)dt,$$

$$w_{23} = \frac{-i}{s} \int_{1/|s|}^{\infty} e^{-ist}(e^{-iht} - 1)F_1(t)dt.$$

Since $1/|s(s+h)| \leq 2/|s|^2$ for $|s| > 2|h|$, it follows from (3.9) that

$$\begin{aligned} \|w_{21}\|_{2,q,H_R} &\leq 3|h||s|^{-2}\|F_0(1/|s|)\|_{2,q,H_R} \\ &\leq C|h||s|^{-2-m+n/2-\varepsilon}(1+|s|)^{-n/2+1+\varepsilon}\|f\|_{q,H} \\ &\leq C|h||s|^{-\beta-1}\|f\|_{q,H}, \end{aligned}$$

and that

$$\begin{aligned} \|w_{22}\|_{2,q,H_R} &\leq 2|h||s|^{-2} \int_{1/|s|}^{\infty} \|F_1(t)\|_{2,q,H_R} dt \\ &\leq C|h||s|^{-2}\|f\|_{q,H} \int_{1/|s|}^{\infty} t^{-1+m-n/2+\varepsilon} dt \\ &\leq C|h||s|^{-\beta-1}\|f\|_{q,H}, \end{aligned}$$

for $f \in L^q_{[d]}(H)$. We perform integration by parts once more to obtain $w_{23} = w_{231} + w_{232} + w_{233}$ with

$$w_{231} = \frac{h}{s^2(s+h)} e^{-i(s+h)/|s|} F_1\left(\frac{1}{|s|}\right) - \frac{1}{s^2} e^{-is/|s|}(e^{-ih/|s|} - 1)F_1\left(\frac{1}{|s|}\right),$$

$$w_{232} = \frac{h}{s^2(s+h)} \int_{1/|s|}^{\infty} e^{-i(s+h)t} F_2(t)dt,$$

$$w_{233} = \frac{-1}{s^2} \int_{1/|s|}^{\infty} e^{-ist}(e^{-iht} - 1)F_2(t)dt.$$

By the same way as in $w_{21} + w_{22}$ we find

$$\begin{aligned} &\|w_{231} + w_{232}\|_{2,q,H_R} \\ \leq\ & 3|h||s|^{-3}\left\{\|F_1(1/|s|)\|_{2,q,H_R} + \int_{1/|s|}^{\infty} \|F_2(t)\|_{2,q,H_R} dt\right\} \\ \leq\ & C|h||s|^{-\beta-1}\|f\|_{q,H}, \end{aligned}$$

for $f \in L^q_{[d]}(H)$. Finally, we use (3.9) again to get

$$\begin{aligned} \|w_{233}\|_{2,q,H_R} &\leq |h||s|^{-2} \int_{1/|s|}^{\infty} t\|F_2(t)\|_{2,q,H_R} dt \\ &\leq C|h||s|^{-2}\|f\|_{q,H} \int_{1/|s|}^{\infty} t^{-1+m-n/2+\varepsilon} dt \\ &\leq C|h||s|^{-\beta-1}\|f\|_{q,H}, \end{aligned}$$

for $f \in L^q_{[d]}(H)$. We gather all the estimates above to conclude (3.7). □

**Remark 3.3.** Estimate (3.7) together with (3.4) implies

$$\int_{-\infty}^{\infty} \|\Phi_H^{(m)}(s+h)f - \Phi_H^{(m)}(s)f\|_{2,q,H_R} ds \le C|h|^{1-\beta}\|f\|_{q,H},$$

for $h \in \boldsymbol{R}$ and $f \in L^q_{[d]}(H)$ (see Lemma 4.4 and its proof), which is related to the assumption of Lemma 5.2. In Lemma 4.4 we will deduce the same regularity of $\partial_s^m(is+A)^{-1}Pf$ for an aperture domain $\Omega$ as above when $f \in L^q(\Omega)$ has a bounded support. For the Oseen resolvent system in the 3-dimensional whole space, Kobayashi and Shibata [40, Lemma 3.6] showed a sharper estimate; indeed, $|h|^{1-\beta}$ can be replaced by $|h|^{1/2}$. Their method is different from ours.

## 4. The Stokes resolvent

In this section, based on the results for the half space obtained in the previous section, we address ourselves to analogous regularity estimates near $\lambda = 0$ of the Stokes resolvent $u = (\lambda + A)^{-1}Pf$, which together with the associated pressure $p$ satisfies the system

$$\lambda u - \Delta u + \nabla p = f, \quad \nabla \cdot u = 0,$$

in an aperture domain $\Omega$ subject to $u|_{\partial\Omega} = 0$ and $\phi(u) = 0$, where $f \in L^q(\Omega), 1 < q < \infty$ and $\lambda \in \boldsymbol{C} \setminus (-\infty, 0]$. To this end, as in [38], [40] and [1], we start with the construction of the resolvent near $\lambda = 0$ for $f \in L^q(\Omega)$ with bounded support. We fix a smooth bounded subdomain $D$ so that $\Omega_{R_0+3} \subset D \subset \Omega$. Given $f \in L^q(\Omega)$, we set $v_0 = A_{q,D}^{-1}P_{q,D}f$ and take a pressure $\pi_0$ associated to $v_0$; they solve the Stokes system

$$-\Delta v_0 + \nabla \pi_0 = f, \quad \nabla \cdot v_0 = 0,$$

in $D$ subject to $v_0|_{\partial D} = 0$, where $f$ is understood as the restriction of $f$ on $D$. We further set

$$v_\pm(x,\lambda) = (\lambda + A_{q,H_\pm})^{-1}P_{q,H_\pm}[\psi_{\pm,R_0}f],$$

where $\psi_{\pm,R_0}$ are the cut-off functions given by (2.1). One needs also the case $\lambda = 0$

$$v_\pm(x,0) = \int_0^\infty E_\pm(t)P_{q,H_\pm}[\psi_{\pm,R_0}f]dt,$$

which is the solution written by the Green tensor for the Stokes problem in $H_\pm$ (see Remark 3.2). We take the pressures $\pi_\pm$ in $H_\pm$ associated to $v_\pm$ so that

$$\int_{D_{\pm,R_0+1}} \{\pi_\pm(x,\lambda) - \pi_0(x)\}dx = 0, \tag{4.1}$$

for each $\lambda$. In this section, for simplicity, we use the abbreviations $\psi_\pm$ for the cut-off functions $\psi_{\pm,R_0+1}$ given by (2.1) and $S_\pm$ for the Bogovskiĭ operators $S_{\pm,R_0+1}$

introduced in section 2. With use of $\{v_\pm, \pi_\pm\}, \{v_0, \pi_0\}$ and $\psi_\pm$ together with $S_\pm$, we set

$$\begin{cases} v &= T(\lambda)f \\ &= \psi_+ v_+ + \psi_- v_- + (1-\psi_+-\psi_-)v_0 \\ &\quad -S_+[(v_+-v_0)\cdot\nabla\psi_+] - S_-[(v_- - v_0)\cdot\nabla\psi_-], \\ \pi &= \psi_+\pi_+ + \psi_-\pi_- + (1-\psi_+-\psi_-)\pi_0. \end{cases} \tag{4.2}$$

We here note that $\int_{D_{\pm,R_0+1}}(v_\pm - v_0)\cdot\nabla\psi_\pm dx = 0$ since $\nabla\cdot v_\pm = \nabla\cdot v_0 = 0$. An elementary calculation shows that the pair $\{v,\pi\}$ satisfies

$$\lambda v - \Delta v + \nabla\pi = f + Q(\lambda)f, \quad \nabla\cdot v = 0, \tag{4.3}$$

in $\Omega$ subject to $v|_{\partial\Omega} = 0$ and

$$\phi(v) = \int_M N\cdot v_0 d\sigma = \int_{\Omega_+\cap D}\nabla\cdot v_0 dx = 0,$$

where

$$Q(\lambda)f = Q_1(\lambda)f + Q_2(\lambda)f \tag{4.4}$$

with

$$\begin{aligned} Q_1(\lambda)f &= \lambda(1-\psi_+-\psi_-)v_0 - 2\nabla\psi_+\cdot\nabla(v_+-v_0) - 2\nabla\psi_-\cdot\nabla(v_- - v_0) \\ &\quad -(\Delta\psi_+)(v_+-v_0) - (\Delta\psi_-)(v_- - v_0) \\ &\quad +(\nabla\psi_+)(\pi_+-\pi_0) + (\nabla\psi_-)(\pi_- - \pi_0) \\ &\quad -\lambda S_+[(v_+-v_0)\cdot\nabla\psi_+] - \lambda S_-[(v_- - v_0)\cdot\nabla\psi_-], \end{aligned}$$

and

$$Q_2(\lambda)f = \Delta S_+[(v_+-v_0)\cdot\nabla\psi_+] + \Delta S_-[(v_- - v_0)\cdot\nabla\psi_-].$$

By (2.2) we have $S_\pm[(v_\pm - v_0)\cdot\nabla\psi_\pm] \in W_0^{2,q}(D_{\pm,R_0+1})$. But one can obtain the regularity of this term only up to $W_0^{2,q}$ (while the $W_0^{3,q}$-regularity of the corresponding term is available for the exterior problem). This is the reason why the remaining term $Q(\lambda)$ has been divided into two parts.

We first derive the regularity estimates near $\lambda = 0$ of $T(\lambda)$ and $Q(\lambda)$.

**Lemma 4.1.** *Let $n \geq 3, 1 < q < \infty, d \geq R_0$ and $R \geq R_0$. For any small $\varepsilon > 0$ there are constants $C_1 = C_1(\Omega, n, q, d, R, \varepsilon) > 0$ and $C_2 = C_2(\Omega, n, q, d, \varepsilon) > 0$ such that*

$$|\lambda|^\beta \|\partial_\lambda^m T(\lambda)f\|_{2,q,\Omega_R} + \sum_{k=0}^{m-1}\|\partial_\lambda^k T(\lambda)f\|_{2,q,\Omega_R} \leq C_1\|f\|_q, \tag{4.5}$$

*for $Re\,\lambda \geq 0$ $(\lambda \neq 0)$ and $f \in L^q_{[d]}(\Omega)$; and*

$$|\lambda|^\beta \|\partial_\lambda^m Q(\lambda)f\|_q + \sum_{k=0}^{m-1}\|\partial_\lambda^k Q(\lambda)f\|_q \leq C_2\|f\|_q, \tag{4.6}$$

*for $Re\,\lambda \geq 0$ with $0 < |\lambda| \leq 2$ and $f \in L^q_{[d]}(\Omega)$, where $m$ and $\beta = \beta(\varepsilon)$ are the same as in Lemma 3.3.*

*Proof.* In view of (4.2), we deduce (4.5) immediately from (3.4) together with (2.2). One can show (4.6) likewise, but it remains to estimate the pressures $\pi_\pm$ contained in (4.4). By (4.1) we have

$$\int_{D_{\pm,R_0+1}} \partial_\lambda^k \pi_\pm(x,\lambda)dx = 0, \quad 1 \le k \le m. \tag{4.7}$$

On the other hand, from the Stokes resolvent system we obtain

$$\lambda\partial_\lambda^k v_\pm + k\partial_\lambda^{k-1} v_\pm - \Delta\partial_\lambda^k v_\pm + \nabla\partial_\lambda^k \pi_\pm = 0, \quad 1 \le k \le m,$$

in $H_\pm$. This combined with (4.7) gives

$$\begin{aligned} &\|(\nabla\psi_\pm)\partial_\lambda^k \pi_\pm(\lambda)\|_q \\ \le\ & C\|\nabla\partial_\lambda^k \pi_\pm(\lambda)\|_{-1,q,D_{\pm,R_0+1}} \\ \le\ & C\|\nabla\partial_\lambda^k v_\pm(\lambda)\|_{q,H_{\pm,R_0+2}} + C|\lambda|\|\partial_\lambda^k v_\pm(\lambda)\|_{q,H_{\pm,R_0+2}} \\ & +Ck\|\partial_\lambda^{k-1} v_\pm(\lambda)\|_{q,H_{\pm,R_0+2}}, \end{aligned}$$

for $1 \le k \le m$. Similarly, for $k = 0$, we use (4.1) to get

$$\begin{aligned} &\|(\nabla\psi_\pm)(\pi_\pm(\lambda) - \pi_0)\|_q \\ \le\ & C\|\nabla(\pi_\pm(\lambda) - \pi_0)\|_{-1,q,D_{\pm,R_0+1}} \\ \le\ & C\|\nabla v_\pm(\lambda)\|_{q,H_{\pm,R_0+2}} + C|\lambda|\|v_\pm(\lambda)\|_{q,H_{\pm,R_0+2}} + C\|f\|_q. \end{aligned}$$

It thus follows from (3.4) that

$$|\lambda|^\beta \|(\nabla\psi_\pm)\partial_\lambda^m \pi_\pm(\lambda)\|_q + \sum_{k=0}^{m-1} \|(\nabla\psi_\pm)\partial_\lambda^k(\pi_\pm(\lambda) - \pi_0)\|_q \le C\|f\|_q,$$

for $\mathrm{Re}\,\lambda \ge 0$ with $0 < |\lambda| \le 2$ and $f \in L^q_{[d]}(\Omega)$. This completes the proof. □

**Remark 4.1.** In the proof above, we have made use of the inequality (see Galdi [28, Chapter III])

$$\|g - \overline{g}\|_{q,G} \le C\|\nabla g\|_{-1,q,G} \quad \text{with } \overline{g} = \frac{1}{|G|}\int_G g(x)dx,$$

for $g \in L^q(G), 1 < q < \infty$, where $G$ is a bounded domain for which the result of Bogovskiĭ [2] introduced in section 2 holds (for instance, $G$ has a locally Lipschitz boundary), although the usual Poincaré inequality leads us to Lemma 4.1 because we have (3.4) in $W^{2,q}(H_R)$. Since the inequality above will be often used later, we give a brief proof for completeness. For each $\varphi \in L^{q/(q-1)}(G)$, we put $\overline{\varphi} = \frac{1}{|G|}\int_G \varphi(x)dx$. Then the result of Bogovskiĭ ensures the existence of $w \in W_0^{1,q/(q-1)}(G)^n$ so that

$$\nabla\cdot w = \varphi - \overline{\varphi}, \quad \|\nabla w\|_{q/(q-1),G} \le C\|\varphi - \overline{\varphi}\|_{q/(q-1),G} \le C\|\varphi\|_{q/(q-1),G}.$$

Therefore,

$$\begin{aligned} |(g - \overline{g}, \varphi)| &= |(g, \varphi - \overline{\varphi})| = |(g, \nabla\cdot w)| = |(-\nabla g, w)| \\ &\le C\|\nabla g\|_{-1,q,G}\|\varphi\|_{q/(q-1),G}, \end{aligned}$$

for all $\varphi \in L^{q/(q-1)}(G)$, which implies the desired inequality.

Let us consider the case $\lambda = 0$ and simply write $v_\pm = v_\pm(x,0)$. Since

$$\|(v_\pm - v_0)\cdot\nabla\psi_\pm\|_{2,q} \le C\|f\|_q,$$

the operator $[f \mapsto (v_\pm - v_0)\cdot\nabla\psi_\pm] : L^q(\Omega) \to W_0^{1,q}(D_{\pm,R_0+1})$ is compact, which combined with (2.2) implies that so is the operator $Q_2(0) : L^q(\Omega) \to L^q_{[d]}(\Omega)$, where $d \ge R_0 + 2$. The other part $Q_1(0)f$ fulfills

$$\|Q_1(0)f\|_{1,q} \le C\|f\|_q,$$

from which the compactness of $Q_1(0) : L^q(\Omega) \to L^q_{[d]}(\Omega)$ follows; as a consequence, $Q(0) = Q_1(0) + Q_2(0)$ is a compact operator from $L^q_{[d]}(\Omega)$, $d \ge R_0 + 2$, into itself. We will show that $1 + Q(0)$ is injective in $L^q_{[d]}(\Omega)$. Let $f \in L^q_{[d]}(\Omega)$ satisfy $(1 + Q(0))f = 0$. In view of (4.3), the pair $\{v, \pi\}$ given by (4.2) for such $f$ should obey

$$-\Delta v + \nabla\pi = 0, \quad \nabla\cdot v = 0,$$

in $\Omega$ subject to $v|_{\partial\Omega} = 0$ and $\phi(v) = 0$. Since $f \in L^r_{[d]}(\Omega)$ for $1 < r < \min\{n, q\}$, we have

$$\nabla^2 v, \nabla\pi \in L^r(\Omega), \quad \nabla v \in L^{nr/(n-r)}(\Omega), \quad v, \pi \in L^r_{loc}(\overline{\Omega});$$

especially, the summability of $\nabla v$ at infinity is implied by the boundedness of the support of $f$. It thus follows from Theorem 1.4 (i) of Farwig [18] that $v = \nabla\pi = 0$; here, it should be remarked that the uniqueness holds without any radiation condition (unlike the exterior problem discussed in [38] and [40]). We go back to (4.2) to see that $v_\pm = \nabla\pi_\pm = f = 0$ in $H_\pm \setminus B_{R_0+2}$ and that $v_0 = \nabla\pi_0 = f = 0$ in $\Omega_{R_0+1}$. Set $U_\pm = (D \cup B_{R_0}) \cap H_\pm$. Both $\{v_\pm, \pi_\pm\}$ and $\{v_0, \pi_0\}$ then belong to $W^{2,q}(U_\pm) \times W^{1,q}(U_\pm)$ and are the solutions of the Stokes system in $U_\pm$ with zero boundary condition for the external force $f$. They thus coincide with each other and, in view of (4.2) again, we have $v_0 = \nabla\pi_0 = f = 0$ in $D$; after all, $f = 0$ in $\Omega$. Owing to the Fredholm theorem, $1 + Q(0)$ has a bounded inverse $(1 + Q(0))^{-1}$ on $L^q_{[d]}(\Omega)$.

Set $\Sigma_\eta = \{\lambda \in \boldsymbol{C}; \mathrm{Re}\,\lambda \ge 0, 0 < |\lambda| \le \eta\}$ for $\eta > 0$. By (4.1) we have $\int_{D_{\pm,R_0+1}}\{\pi_\pm(x,\lambda) - \pi_\pm(x,0)\}dx = 0$, which together with the Stokes system in $H_\pm$ yields

$$\begin{aligned}
&\|(\nabla\psi_\pm)(\pi_\pm(\lambda) - \pi_\pm(0))\|_q \\
\le\ & C\|\nabla(\pi_\pm(\lambda) - \pi_\pm(0))\|_{-1,q,D_{\pm,R_0+1}} \\
\le\ & C\|\nabla(v_\pm(\lambda) - v_\pm(0))\|_{q,H_{\pm,R_0+2}} + C|\lambda|\|v_\pm(\lambda)\|_{q,H_{\pm,R_0+2}}.
\end{aligned}$$

Therefore, we get

$$\begin{aligned}
&\|Q(\lambda)f - Q(0)f\|_q \\
\le\ & C\|v_+(\lambda) - v_+(0)\|_{1,q,H_{+,R_0+2}} + C\|v_-(\lambda) - v_-(0)\|_{1,q,H_{-,R_0+2}} \\
&+ C|\lambda|\{\|v_+(\lambda)\|_{q,H_{+,R_0+2}} + \|v_-(\lambda)\|_{q,H_{-,R_0+2}} + \|v_0\|_{q,D}\}.
\end{aligned}$$

From (3.5) we thus obtain

$$\|Q(\lambda) - Q(0)\|_{B(L^q_{[d]}(\Omega))} \to 0,$$

as $\lambda \to 0$ with $\mathrm{Re}\,\lambda \geq 0$, which implies the existence of a constant $\eta > 0$ such that $1 + Q(\lambda)$ has also a bounded inverse (in terms of the Neumann series) on $L^q_{[d]}(\Omega)$ with uniform bounds

$$\|(1 + Q(\lambda))^{-1}\|_{B(L^q_{[d]}(\Omega))} \leq C, \tag{4.8}$$

for $\lambda \in \Sigma_\eta \cup \{0\}$. Since the resolvent is uniquely determined, one can represent it for $\lambda \in \Sigma_\eta$ and $f \in L^q_{[d]}(\Omega), d \geq R_0 + 2$, as

$$(\lambda + A)^{-1}Pf = T(\lambda)(1 + Q(\lambda))^{-1}f. \tag{4.9}$$

We are in a position to show an analogous result for the resolvent to (3.4).

**Lemma 4.2.** *Let $n \geq 3, 1 < q < \infty, d \geq R_0$ and $R \geq R_0$. Given $f \in L^q_{[d]}(\Omega)$, set $u(\lambda) = (\lambda + A)^{-1}Pf$. For any small $\varepsilon > 0$ there is a constant $C = C(\Omega, n, q, d, R, \varepsilon) > 0$ such that*

$$|\lambda|^\beta \|\partial_\lambda^m u(\lambda)\|_{2,q,\Omega_R} + \sum_{k=0}^{m-1} \|\partial_\lambda^k u(\lambda)\|_{2,q,\Omega_R} \leq C\|f\|_q, \tag{4.10}$$

*for $\mathrm{Re}\,\lambda \geq 0$ $(\lambda \neq 0)$ and $f \in L^q_{[d]}(\Omega)$, where $m$ and $\beta = \beta(\varepsilon)$ are the same as in Lemma 3.3.*

*Proof.* The problem is only near $\lambda = 0$ because we have (2.3) for $G = \Omega$. We may also assume $d \geq R_0 + 2$ since $L^q_{[R_0]}(\Omega) \subset L^q_{[d]}(\Omega)$ for such $d$. It thus suffices to show (4.10) for $\lambda \in \Sigma_\eta$ by use of (4.9). For such $\lambda$ and $0 \leq k \leq m$ we see that $\partial_\lambda^k(1 + Q(\lambda))^{-1} \in B(L^q_{[d]}(\Omega))$; furthermore,

$$|\lambda|^\beta \|\partial_\lambda^m (1 + Q(\lambda))^{-1}f\|_q + \sum_{k=0}^{m-1} \|\partial_\lambda^k(1 + Q(\lambda))^{-1}f\|_q \leq C\|f\|_q, \tag{4.11}$$

for $f \in L^q_{[d]}(\Omega)$. In fact, we have the representation

$$\begin{aligned} &\partial_\lambda^k(1 + Q(\lambda))^{-1}f \\ = \;& -(1 + Q(\lambda))^{-1}[\partial_\lambda^k Q(\lambda)](1 + Q(\lambda))^{-1}f + L_k(\lambda)(1 + Q(\lambda))^{-1}f, \end{aligned} \tag{4.12}$$

for $k \geq 1$ and $f \in L^q_{[d]}(\Omega)$, where $L_1(\lambda) = 0$ and $L_k(\lambda)$ with $k \geq 2$ consists of finite sums of finite products of $(1+Q(\lambda))^{-1}, \partial_\lambda Q(\lambda), \cdots, \partial_\lambda^{k-1}Q(\lambda)$. Consequently, (4.6) together with (4.8) implies (4.11). In view of

$$\partial_\lambda^k u(\lambda) = \sum_{j=0}^{k} \binom{k}{j} \partial_\lambda^{k-j} T(\lambda)\, \partial_\lambda^j (1 + Q(\lambda))^{-1}f,$$

we conclude (4.10) from (4.5) and (4.11). □

In the last part of this section we will complete the regularity estimate of the resolvent. To this end, we employ Lemma 3.4 to show the following lemma.

**Lemma 4.3.** *Let $n \geq 3, 1 < q < \infty, d \geq R_0$ and $R \geq R_0$. Set*

$$T^{(k)}(s) = \partial_s^k T(is), \quad Q^{(k)}(s) = \partial_s^k Q(is) \quad (s \in \boldsymbol{R} \setminus \{0\},\ 0 \leq k \leq m).$$

*For any small $\varepsilon > 0$ there is a constant $C = C(\Omega, n, q, d, R, \varepsilon) > 0$ such that*

$$\|T^{(k)}(s+h)f - T^{(k)}(s)f\|_{2,q,\Omega_R} + \|Q^{(k)}(s+h)f - Q^{(k)}(s)f\|_q$$
$$\leq \begin{cases} C|h||s|^{-\beta-1}\|f\|_q & \text{if } k = m, \\ C|h||s|^{-\beta}\|f\|_q & \text{if } k = m-1, \\ C|h|\|f\|_q & \text{if } n \geq 5,\ 0 \leq k \leq m-2, \end{cases} \tag{4.13}$$

*for $2|h| < |s| \leq 1$ and $f \in L^q_{[d]}(\Omega)$, where $m$ and $\beta = \beta(\varepsilon)$ are the same as in Lemma 3.3. Concerning the first term of the left-hand side, (4.13) holds true for $h \in \boldsymbol{R}$ and $|s| > 2|h|$.*

*Proof.* Set

$$v_\pm^{(k)}(s) = \partial_s^k v_\pm(is), \quad \pi_\pm^{(k)}(s) = \partial_s^k \pi_\pm(is) \quad (s \in \boldsymbol{R} \setminus \{0\},\ k = m \text{ or } m-1).$$

It then follows from (4.2) together with (2.2) that

$$\begin{aligned} &\|T^{(m)}(s+h)f - T^{(m)}(s)f\|_{2,q,\Omega_R} \\ \leq\ & C\|v_+^{(m)}(s+h) - v_+^{(m)}(s)\|_{2,q,H_{+,R}} + C\|v_-^{(m)}(s+h) - v_-^{(m)}(s)\|_{2,q,H_{-,R}}. \end{aligned}$$

In order to estimate $Q^{(m)}$, let us investigate the pressures $\pi_\pm^{(m)}$. Similarly to the proof of Lemma 4.1 with the aid of (4.7), one can show

$$\begin{aligned} &\|(\nabla\psi_\pm)\{\pi_\pm^{(m)}(s+h) - \pi_\pm^{(m)}(s)\}\|_q \\ \leq\ & C\|\nabla\pi_\pm^{(m)}(s+h) - \nabla\pi_\pm^{(m)}(s)\|_{-1,q,D_{\pm,R_0+1}} \\ \leq\ & C\|\nabla v_\pm^{(m)}(s+h) - \nabla v_\pm^{(m)}(s)\|_{q,H_{\pm,R_0+2}} \\ & +C\|(s+h)v_\pm^{(m)}(s+h) - sv_\pm^{(m)}(s)\|_{q,H_{\pm,R_0+2}} \\ & +Cm\|v_\pm^{(m-1)}(s+h) - v_\pm^{(m-1)}(s)\|_{q,H_{\pm,R_0+2}}. \end{aligned}$$

This combined with estimates on the other terms by use of (2.2) yields

$$\begin{aligned} &\|Q^{(m)}(s+h)f - Q^{(m)}(s)f\|_q \\ \leq\ & C\|v_+^{(m)}(s+h) - v_+^{(m)}(s)\|_{1,q,H_{+,R_0+2}} \\ & +C\|v_-^{(m)}(s+h) - v_-^{(m)}(s)\|_{1,q,H_{-,R_0+2}} \\ & +C|s|\|v_+^{(m)}(s+h) - v_+^{(m)}(s)\|_{q,H_{+,R_0+2}} \\ & +C|s|\|v_-^{(m)}(s+h) - v_-^{(m)}(s)\|_{q,H_{-,R_0+2}} \\ & +C|h|\|v_+^{(m)}(s+h)\|_{q,H_{+,R_0+2}} + C|h|\|v_-^{(m)}(s+h)\|_{q,H_{-,R_0+2}} \\ & +Cm\|v_+^{(m-1)}(s+h) - v_+^{(m-1)}(s)\|_{q,H_{+,R_0+2}} \\ & +Cm\|v_-^{(m-1)}(s+h) - v_-^{(m-1)}(s)\|_{q,H_{-,R_0+2}}. \end{aligned}$$

Hence (3.7), (3.8) and (3.4) imply (4.13) for the case $k = m$. For $0 \leq k \leq m-1$ we have

$$\|T^{(k)}(s+h)f - T^{(k)}(s)f\|_{2,q,\Omega_R} \leq \left| \int_s^{s+h} \|T^{(k+1)}(\tau)f\|_{2,q,\Omega_R} d\tau \right|,$$

$$\|Q^{(k)}(s+h)f - Q^{(k)}(s)f\|_q \le \left| \int_s^{s+h} \|Q^{(k+1)}(\tau)f\|_q d\tau \right|,$$

which together with (4.5) and (4.6) respectively lead us to (4.13). The proof is thus complete. □

The regularity of the resolvent along the imaginary axis given by the following lemma plays a crucial role in the next section.

**Lemma 4.4.** *Let $n \ge 3, 1 < q < \infty, d \ge R_0$ and $R \ge R_0$. Set*

$$\Phi^{(m)}(s) = \partial_s^m (is + A)^{-1} P \qquad (s \in \boldsymbol{R} \setminus \{0\}).$$

*For any small $\varepsilon > 0$ there is a constant $C = C(\Omega, n, q, d, R, \varepsilon) > 0$ such that*

$$\int_{-\infty}^{\infty} \|\Phi^{(m)}(s+h)f - \Phi^{(m)}(s)f\|_{2,q,\Omega_R} ds \le C|h|^{1-\beta}\|f\|_q, \tag{4.14}$$

*for $|h| < h_0 = \min\{\eta/4, 1/2\}$ and $f \in L^q_{[d]}(\Omega)$. Here, $m$ and $\beta = \beta(\varepsilon)$ are the same as in Lemma 3.3, and $\eta > 0$ is the constant such that (4.9) is valid for $\lambda \in \Sigma_\eta$.*

*Proof.* We may assume $d \ge R_0 + 2$ (as in the proof of Lemma 4.2). Given $h$ satisfying $|h| < h_0$, we divide the integral into three parts

$$\int_{-\infty}^{\infty} \|\Phi^{(m)}(s+h)f - \Phi^{(m)}(s)f\|_{2,q,\Omega_R} ds$$

$$= \int_{|s| \le 2|h|} + \int_{2|h| < |s| \le 2h_0} + \int_{|s| > 2h_0} = I_1 + I_2 + I_3.$$

With the aid of (4.10), we find

$$I_1 \le 2 \int_{|s| \le 3|h|} \|\Phi^{(m)}(s)f\|_{2,q,\Omega_R} ds \le C|h|^{1-\beta}\|f\|_q,$$

for $f \in L^q_{[d]}(\Omega)$. In order to estimate $I_2$, we use the representation

$$\Phi^{(m)}(s)f = \sum_{j=0}^{m} \binom{m}{j} T^{(m-j)}(s)\, V^{(j)}(s)f,$$

where

$$V^{(j)}(s) = \partial_s^j (1 + Q(is))^{-1} \in B(L^q_{[d]}(\Omega)) \qquad (0 < |s| \le \eta,\ 0 \le j \le m).$$

Then,

$$\begin{aligned} &\Phi^{(m)}(s+h)f - \Phi^{(m)}(s)f \\ = &\sum_{j=0}^{m} \binom{m}{j} [T^{(m-j)}(s+h) - T^{(m-j)}(s)]\, V^{(j)}(s+h)f \\ &+ \sum_{j=0}^{m} \binom{m}{j} T^{(m-j)}(s)\, [V^{(j)}(s+h) - V^{(j)}(s)]f. \end{aligned}$$

We first show

$$\|V^{(j)}(s+h)f - V^{(j)}(s)f\|_q$$
$$\leq \begin{cases} C|h||s|^{-\beta-1}\|f\|_q & \text{if } j = m, \\ C|h||s|^{-\beta}\|f\|_q & \text{if } j = m-1, \\ C|h|\|f\|_q & \text{if } n \geq 5,\ 0 \leq j \leq m-2, \end{cases} \tag{4.15}$$

for $2|h| < |s| \leq 2h_0$ and $f \in L^q_{[d]}(\Omega)$. Similarly to the proof of (4.13) for $0 \leq k \leq m-1$, (4.11) implies (4.15) for $0 \leq j \leq m-1$. As in (4.12), we have

$$V^{(m)}(s) = -V^{(0)}(s)Q^{(m)}(s)V^{(0)}(s) + W_m(s)V^{(0)}(s),$$

where $W_1(s) = 0$ and, for $m \geq 2$, $W_m(s) = i^m L_m(is)$ consists of finite sums of finite products of $V^0(s), Q^{(1)}(s), \cdots, Q^{(m-1)}(s)$. Therefore, we collect (4.6), (4.8), (4.13) and (4.15) for $j = 0$ to arrive at (4.15) for $j = m$. It thus follows from (4.5), (4.11), (4.13) and (4.15) that

$$\|\Phi^{(m)}(s+h)f - \Phi^{(m)}(s)f\|_{2,q,\Omega_R} \leq C|h||s|^{-\beta-1}\|f\|_q,$$

for $2|h| < |s| \leq 2h_0$ and $f \in L^q_{[d]}(\Omega)$. As a consequence, we are led to

$$I_2 \leq C|h|\|f\|_q \int_{|s|>2|h|} |s|^{-\beta-1} ds \leq C|h|^{1-\beta}\|f\|_q,$$

for $f \in L^q_{[d]}(\Omega)$. Finally, to estimate $I_3$, one does not need any localization. In fact, since

$$\Phi^{(m)}(s+h)f - \Phi^{(m)}(s)f = (-i)^{m+1}(m+1)! \int_s^{s+h} (i\tau + A)^{-(m+2)} Pf d\tau,$$

(2.3) gives

$$\begin{aligned} \|\Phi^{(m)}(s+h)f - \Phi^{(m)}(s)f\|_{2,q,\Omega_R} &\leq C\|\Phi^{(m)}(s+h)f - \Phi^{(m)}(s)f\|_{D(A_q)} \\ &\leq C|h||s|^{-(m+1)}\|f\|_q, \end{aligned}$$

for $|s| > 2h_0$ ($> 2|h|$) and $f \in L^q(\Omega)$. Therefore, we obtain

$$I_3 \leq C|h|\|f\|_q \int_{|s|>2h_0} |s|^{-(m+1)} ds \leq C|h|\|f\|_q,$$

for $f \in L^q(\Omega)$. Collecting the estimates above on $I_1, I_2$ and $I_3$, we conclude (4.14). □

## 5. $L^q$-$L^r$ estimates of the Stokes semigroup

In this section we will prove Theorem 2.1. As explained in Section 1, the first step is to derive (1.6) for non-solenoidal data with bounded support.

**Lemma 5.1.** *Let $n \geq 3, 1 < q < \infty, d \geq R_0$ and $R \geq R_0$. For any small $\varepsilon > 0$ there is a constant $C = C(\Omega, n, q, d, R, \varepsilon) > 0$ such that*

$$\|e^{-tA}Pf\|_{1,q,\Omega_R} \leq Ct^{-1/2}(1+t)^{-n/2+1/2+\varepsilon}\|f\|_q, \tag{5.1}$$

*for $t > 0$ and $f \in L^q_{[d]}(\Omega)$.*

For the proof, the following lemma due to Shibata is crucial since we know the regularity of the Stokes resolvent given by Lemmas 4.2 and 4.4.

**Lemma 5.2.** *Let $X$ be a Banach space with norm $\|\cdot\|$ and $g \in L^1(\boldsymbol{R}; X)$. If there are constants $\theta \in (0,1)$ and $M > 0$ such that*

$$\int_{-\infty}^{\infty} \|g(s)\| ds + \sup_{h\neq 0} \frac{1}{|h|^\theta} \int_{-\infty}^{\infty} \|g(s+h) - g(s)\| ds \leq M,$$

*then the Fourier inverse image*

$$G(t) = \frac{1}{2\pi} \int_{-\infty}^{\infty} e^{ist} g(s) ds$$

*of $g$ enjoys*

$$\|G(t)\| \leq CM(1+|t|)^{-\theta},$$

*with some $C > 0$ independent of $t \in \boldsymbol{R}$.*

**Remark 5.1.** The assumption of Lemma 5.2 is equivalent to

$$g \in \left(L^1(\boldsymbol{R}; X),\ W^{1,1}(\boldsymbol{R}; X)\right)_{\theta,\infty},$$

where $(\cdot,\cdot)_{\theta,\infty}$ denotes the real interpolation functor (the space to which $g$ belongs is known as a Besov space).

*Proof of Lemma* 5.2. Although this lemma was already proved by Shibata [56], we give our different proof which seems to be simpler. Since $\|G(t)\| \leq M/2\pi$, it suffices to consider the case $|t| > 1$. It is easily seen that if $ht \neq 2j\pi$ $(j = 0, \pm 1, \pm 2, \ldots)$, then

$$G(t) = \frac{e^{iht}}{2\pi(1-e^{iht})} \int_{-\infty}^{\infty} e^{ist}(g(s+h) - g(s)) ds,$$

from which the assumption leads us to

$$\|G(t)\| \leq \frac{M|h|^\theta}{2\pi|1-e^{iht}|}.$$

Taking $h = 1/t$ immediately implies the desired estimate. □

*Proof of Lemma* 5.1. Since

$$\|e^{-tA}Pf\|_{1,q} \leq C\|e^{-tA}Pf\|_{D(A_q)}^{1/2} \|e^{-tA}Pf\|_q^{1/2} \leq Ct^{-1/2}\|f\|_q, \tag{5.2}$$

for $0 < t < 1$ and $f \in L^q(\Omega)$, we will concentrate ourselves on the proof of (5.1) for $t \geq 1$, namely (1.6). Given $R \geq R_0$, we set $\psi = 1 - \psi_{+,R} - \psi_{-,R}$, where the cut-off functions $\psi_{\pm,R}$ are given by (2.1). One can justify the following representation formula of the semigroup for $f \in L^q_{[d]}(\Omega)$:

$$\psi e^{-tA}Pf = \frac{i^m}{2\pi t^m} \int_{-\infty}^{\infty} e^{ist} \psi \Phi^{(m)}(s) f ds, \tag{5.3}$$

where $\Phi^{(m)}(s) = \partial_s^m (is + A)^{-1}P$ and $m$ is the same as in Lemma 3.3. In fact, starting from the standard Dunford integral representation, we perform $m$-times integrations by parts and then move the path of integration to the imaginary axis but avoid the origin $\lambda = 0$, so that

$$\psi e^{-tA}Pf = \frac{i^m}{2\pi t^m}\left\{\int_{-\infty}^{-\delta} + \int_{\delta}^{\infty}\right\} e^{ist}\psi\Phi^{(m)}(s)fds + \frac{(-1)^m}{2\pi i t^m}\int_{\Gamma_\delta} e^{\lambda t}\psi\partial_\lambda^m(\lambda + A)^{-1}Pfd\lambda,$$

for any $\delta > 0$, where $\Gamma_\delta = \{\delta e^{i\theta}; -\pi/2 \le \theta \le \pi/2\}$ (this formula is valid for $f \in L^q(\Omega)$ without $\psi$). Owing to (4.10), the last integral vanishes in $L^q(\Omega)$ as $\delta \to 0$ for $f \in L^q_{[d]}(\Omega)$; thus, we arrive at (5.3). Now, it follows from (4.10) and $\|\Phi^{(m)}(s)f\|_{1,q} \le C\|\Phi^{(m)}(s)f\|_{D(A_q)}^{1/2}\|\Phi^{(m)}(s)f\|_q^{1/2}$ together with (2.3) that

$$\int_{-\infty}^{\infty} \|\psi\Phi^{(m)}(s)f\|_{1,q}ds \le C\int_{|s|\le 1}\frac{\|f\|_q}{|s|^\beta}ds + C\int_{|s|>1}\frac{\|f\|_q}{|s|^{m+1/2}}ds \le C\|f\|_q.$$

Further, (4.14) and the estimate above respectively imply that

$$\sup_{0<|h|<h_0}\frac{1}{|h|^{1-\beta}}\int_{-\infty}^{\infty} \|\psi\Phi^{(m)}(s+h)f - \psi\Phi^{(m)}(s)f\|_{1,q}ds \le C\|f\|_q,$$

and that

$$\begin{aligned}\sup_{|h|\ge h_0}\frac{1}{|h|^{1-\beta}}&\int_{-\infty}^{\infty} \|\psi\Phi^{(m)}(s+h)f - \psi\Phi^{(m)}(s)f\|_{1,q}ds \\ &\le \frac{2}{h_0^{1-\beta}}\int_{-\infty}^{\infty}\|\psi\Phi^{(m)}(s)f\|_{1,q}ds \le C\|f\|_q.\end{aligned}$$

Hence, we can apply Lemma 5.2 with $X = W^{1,q}(\Omega)$ and $g(s) = \psi\Phi^{(m)}(s)f$ to the formula (5.3); as a consequence, we obtain

$$\|e^{-tA}Pf\|_{1,q,\Omega_R} \le \|\psi e^{-tA}Pf\|_{1,q} \le Ct^{-m}(1+t)^{-1+\beta}\|f\|_q,$$

for $t > 0$, which implies (5.1) for $t \ge 1$ and $f \in L^q_{[d]}(\Omega)$. This completes the proof. □

**Remark 5.2.** It is possible to show the decay rate $t^{-n/2+\varepsilon}$ of the semigroup in $W^{2,q}(\Omega_R)$ as well. This follows immediately from the proof given above with $X = W^{2,q}(\Omega)$ for $n \ge 5$. When $n = 3$ or 4 (thus $m = 1$), as in Kobayashi and Shibata [40], we have to introduce a cut-off function $\rho \in C_0^\infty(\boldsymbol{R}; [0,1])$ with $\rho(s) = 1$ near $s = 0$; then one can employ Lemma 5.2 with $X = W^{2,q}(\Omega)$ and $g(s) = \rho(s)\psi\Phi^{(m)}(s)f$ to obtain the desired result since a rapid decay of the remaining integral far from $s = 0$ is derived via integration by parts. We did not follow this procedure because Lemma 5.1 is sufficient for the proof of Theorem 2.1.

The next step is to deduce the sharp local energy decay estimate (1.5) from Lemma 5.1.

**Lemma 5.3.** *Let $n \geq 3, 1 < q < \infty$ and $R \geq R_0$. Then there is a constant $C = C(\Omega, n, q, R) > 0$ such that*

$$\|e^{-tA}f\|_{1,q,\Omega_R} \leq Ct^{-n/2q}\|f\|_q, \tag{5.4}$$

*for $t \geq 2$ and $f \in L^q_\sigma(\Omega)$; and*

$$\|e^{-tA}f\|_{1,q,\Omega_R} + \|\partial_t e^{-tA}f\|_{q,\Omega_R} \leq C(1+t)^{-n/2q}\|f\|_{D(A_q)}, \tag{5.5}$$

*for $t \geq 0$ and $f \in D(A_q)$.*

*Proof.* We employ a localization procedure which is similar to [38] and [40]. Given $f \in L^q_\sigma(\Omega)$, we set $g = e^{-A}f \in D(A_q)$ and intend to derive the decay estimate of $u(t) = e^{-tA}g = e^{-(t+1)A}f$ in $W^{1,q}(\Omega_R)$ for $t \geq 1$. We denote by $p$ the pressure associated to $u$. We make use of the cut-off functions given by (2.1) and the Bogovskiĭ operator introduced in section 2. Set

$$g_\pm = \psi_{\pm,R_0+1}\, g - S_{\pm,R_0+1}[g \cdot \nabla\psi_{\pm,R_0+1}],$$

and

$$v_\pm(t) = E_\pm(t)g_\pm.$$

Note that $\int_{D_{\pm,R_0+1}} g \cdot \nabla\psi_{\pm,R_0+1}dx = 0$ and that $g_\pm \in D(A_{q,H_\pm})$ with

$$\|g_\pm\|_{D(A_{q,H_\pm})} \leq C\|g_\pm\|_{2,q,H_\pm} \leq C\|g\|_{2,q} \leq C\|g\|_{D(A_q)} \leq C\|f\|_q, \tag{5.6}$$

by (2.2). We take the pressures $\pi_\pm$ in $H_\pm$ associated to $v_\pm$ in such a way that

$$\int_{D_{\pm,R_0}} \pi_\pm(x,t)dx = 0, \tag{5.7}$$

for each $t$. In the course of the proof of this lemma, for simplicity, we abbreviate $\psi_{\pm,R_0}$ to $\psi_\pm$ and $S_{\pm,R_0}$ to $S_\pm$. We now define $\{u_\pm, p_\pm\}$ by

$$u_\pm(t) = \psi_\pm v_\pm(t) - S_\pm[v_\pm(t) \cdot \nabla\psi_\pm], \quad p_\pm(t) = \psi_\pm\pi_\pm(t).$$

Then it follows from Lemma 3.2 together with (2.2) and (5.6) that

$$\|u_\pm(t)\|_{1,q,\Omega_R} \leq C\|v_\pm(t)\|_{1,q,H_{\pm,L}} \leq C(1+t)^{-n/2q}\|f\|_q, \tag{5.8}$$

for $t \geq 0$, where $L = \max\{R, R_0+1\}$. Thus, in order to estimate $u(t)$, let us consider

$$v(t) = u(t) - u_+(t) - u_-(t), \quad \pi(t) = p(t) - p_+(t) - p_-(t),$$

which should obey

$$\partial_t v - \Delta v + \nabla\pi = K, \quad \nabla \cdot v = 0,$$

in $\Omega$ subject to $v|_{\partial\Omega} = 0,\ \phi(v) = \phi(u) = 0$ and

$$v|_{t=0} = v_0 = g - g_+ - g_- \in L^q_{[R_0+2]}(\Omega) \cap D(A_q),$$

where

$$\begin{aligned} K = \ & 2\nabla\psi_+ \cdot \nabla v_+ + 2\nabla\psi_- \cdot \nabla v_- + (\Delta\psi_+)v_+ + (\Delta\psi_-)v_- \\ & -\Delta S_+[v_+ \cdot \nabla\psi_+] - \Delta S_-[v_- \cdot \nabla\psi_-] \\ & +S_+[\partial_t v_+ \cdot \nabla\psi_+] + S_-[\partial_t v_- \cdot \nabla\psi_-] - (\nabla\psi_+)\pi_+ - (\nabla\psi_-)\pi_-; \end{aligned}$$

we here note that $\nabla \cdot K \neq 0$ as well as $K|_{\partial\Omega} \neq 0$ and we can obtain the regularity of $K$ only up to $L^q$ (in contrast to the exterior problem discussed in [38] and [40]). By (5.7) and in view of the Stokes system in $H_\pm$ we have

$$\begin{aligned}\|(\nabla\psi_\pm)\pi_\pm(t)\|_q &\leq C\|\nabla\pi_\pm(t)\|_{-1,q,D_{\pm,R_0}} \\ &\leq C\|\nabla v_\pm(t)\|_{q,H_{\pm,R_0+1}} + C\|\partial_t v_\pm(t)\|_{q,H_{\pm,R_0+1}},\end{aligned}$$

which together with (2.2) implies $K(t) \in L^q_{[R_0+1]}(\Omega)$ and

$$\begin{aligned}\|K(t)\|_q \leq \; & C\|v_+(t)\|_{1,q,H_{+,R_0+1}} + C\|v_-(t)\|_{1,q,H_{-,R_0+1}} \\ & +C\|\partial_t v_+(t)\|_{q,H_{+,R_0+1}} + C\|\partial_t v_-(t)\|_{q,H_{-,R_0+1}}.\end{aligned}$$

Therefore, Lemma 3.2 and (5.6) yield

$$\|K(t)\|_q \leq C(1+t)^{-n/2q}\|f\|_q, \tag{5.9}$$

for $t \geq 0$. In order to estimate

$$v(t) = e^{-tA}v_0 + \int_0^t e^{-(t-\tau)A}PK(\tau)d\tau,$$

we employ Lemma 5.1. By (5.1) with a suitable $\varepsilon > 0$ and (5.6) we find

$$\|e^{-tA}v_0\|_{1,q,\Omega_R} \leq Ct^{-n/2+\varepsilon}\|v_0\|_q \leq Ct^{-n/2q}\|f\|_q,$$

for $t \geq 1$. We next combine (5.1) with (5.9) to get

$$\begin{aligned}& \int_0^t \|e^{-(t-\tau)A}PK(\tau)\|_{1,q,\Omega_R}d\tau \\ \leq \;& C\|f\|_q \int_0^t (t-\tau)^{-1/2}(1+t-\tau)^{-n/2+1/2+\varepsilon}(1+\tau)^{-n/2q}d\tau \\ = \;& C\|f\|_q(I_1+I_2),\end{aligned}$$

where $I_1 = \int_0^{t/2}$ and $I_2 = \int_{t/2}^t$. An elementary calculation gives

$$I_1 \leq \left\{\begin{array}{ll} Ct^{-1/2}(1+t/2)^{-n/2-n/2q+3/2+\varepsilon} & \text{if } q > n/2 \\ Ct^{-1/2}(1+t/2)^{-n/2+1/2+\varepsilon}\log(1+t/2) & \text{if } q = n/2 \\ Ct^{-1/2}(1+t/2)^{-n/2+1/2+\varepsilon} & \text{if } q < n/2 \end{array}\right\} \leq Ct^{-n/2q},$$

for $t \geq 1$ and

$$I_2 \leq (1+t/2)^{-n/2q}\int_0^\infty \tau^{-1/2}(1+\tau)^{-n/2+1/2+\varepsilon}d\tau \leq C(1+t/2)^{-n/2q},$$

for $t > 0$. We collect the estimates above to obtain

$$\|v(t)\|_{1,q,\Omega_R} \leq Ct^{-n/2q}\|f\|_q, \tag{5.10}$$

for $t \geq 1$. From (5.8) and (5.10) we deduce

$$\|u(t)\|_{1,q,\Omega_R} = \|v(t) + u_+(t) + u_-(t)\|_{1,q,\Omega_R} \leq Ct^{-n/2q}\|f\|_q,$$

for $t \geq 1$ and $f \in L^q_\sigma(\Omega)$, which proves (5.4). Let $f \in D(A_q)$. Then we easily observe

$$\|e^{-tA}f\|_{1,q,\Omega_R} + \|\partial_t e^{-tA}f\|_{q,\Omega_R} \leq C\|e^{-tA}f\|_{D(A_q)} \leq C\|f\|_{D(A_q)},$$

for $t \geq 0$ and also we can estimate $\partial_t e^{-tA} f$ for large $t$; in fact, by virtue of (5.4) just proved we get

$$\|\partial_t e^{-tA} f\|_{q,\Omega_R} = \|e^{-tA} Af\|_{q,\Omega_R} \leq Ct^{-n/2q}\|Af\|_q,$$

for $t \geq 2$. This implies (5.5). □

We are interested in the $L^q$ estimate of $\nabla e^{-tA}$ for large $t$, in particular, the $L^n$ estimate is quite important for us.

**Lemma 5.4.** *Let $n \geq 3$ and $1 < q < \infty$. Then there is a constant $C = C(\Omega, n, q) > 0$ such that*

$$\|\nabla e^{-tA} f\|_q \leq Ct^{-\min\{1/2, n/2q\}}\|f\|_q, \tag{5.11}$$

*for $t \geq 2$ and $f \in L^q_\sigma(\Omega)$.*

*Proof.* We fix $R \geq R_0 + 1$. Since we have already known the decay rate $t^{-n/2q}$ of $\|\nabla e^{-tA} f\|_{q,\Omega_R}$ by Lemma 5.3, it suffices to derive the estimate outside $\Omega_R$, that is,

$$\|\nabla e^{-tA} f\|_{q,\Omega_\pm\setminus\Omega_R} \leq Ct^{-\min\{1/2, n/2q\}}\|f\|_q, \tag{5.12}$$

for $t \geq 2$ and $f \in L^q_\sigma(\Omega)$. In an analogous way to [38], [40] and [1], we make use of the decay properties of the semigroup $E_\pm(t)$ for the half space. Given $f \in L^q_\sigma(\Omega)$, we set $g = e^{-A} f \in D(A_q)$ and then $u(t) = e^{-tA} g = e^{-(t+1)A} f$. We choose two pressures $p_\pm$ in $\Omega$ associated to $u$ in such a way that

$$\int_{D_{\pm,R-1}} p_\pm(x,t)dx = 0, \tag{5.13}$$

for each $t$ ($p_+$ and $p_-$ will be used independently). With use of the cut-off functions given by (2.1) and the Bogovskiĭ operator introduced in section 2, we define $\{v_\pm, \pi_\pm\}$ by

$$v_\pm(t) = \psi_\pm u(t) - S_\pm[u(t)\cdot\nabla\psi_\pm], \quad \pi_\pm(t) = \psi_\pm p_\pm(t).$$

Here and in what follows, we use the abbreviations $\psi_\pm$ for $\psi_{\pm,R-1}$ and $S_\pm$ for $S_{\pm,R-1}$. Since $v_\pm = u$ for $x \in \Omega_\pm \setminus \Omega_R = H_\pm \setminus B_R$, we will show

$$\|\nabla v_\pm(t)\|_{q,H_\pm} \leq Ct^{-\min\{1/2, n/2q\}}\|g\|_{D(A_q)}, \tag{5.14}$$

for $t \geq 1$, which combined with $\|g\|_{D(A_q)} \leq C\|f\|_q$ implies (5.12) for $t \geq 2$. It is easily observed that $\{v_\pm, \pi_\pm\}$ satisfies

$$\partial_t v_\pm - \Delta v_\pm + \nabla\pi_\pm = Z_\pm, \quad \nabla\cdot v_\pm = 0,$$

in $H_\pm$ subject to $v_\pm|_{\partial H_\pm} = 0$ and

$$v_\pm|_{t=0} = a_\pm = \psi_\pm g - S_\pm[g\cdot\nabla\psi_\pm],$$

where

$$\begin{aligned} Z_\pm = \; & -2\nabla\psi_\pm\cdot\nabla u - (\Delta\psi_\pm)u + \Delta S_\pm[u\cdot\nabla\psi_\pm] \\ & -S_\pm[\partial_t u\cdot\nabla\psi_\pm] + (\nabla\psi_\pm)p_\pm. \end{aligned}$$

Our task is now to estimate the gradient of

$$v_\pm(t) = E_\pm(t)a_\pm + \int_0^t E_\pm(t-\tau)P_{H_\pm}Z_\pm(\tau)d\tau. \tag{5.15}$$

By virtue of (5.13) we have

$$\begin{aligned}\|(\nabla\psi_\pm)p_\pm(t)\|_{q,H_\pm} &\le C\|\nabla p_\pm(t)\|_{-1,q,D_{\pm,R-1}} \\ &\le C\|\nabla u(t)\|_{q,\Omega_R} + C\|\partial_t u(t)\|_{q,\Omega_R},\end{aligned}$$

from which together with (2.2) it follows that

$$\|Z_\pm(t)\|_{q,H_\pm} \le C\|u(t)\|_{1,q,\Omega_R} + C\|\partial_t u(t)\|_{q,\Omega_R}.$$

Hence, (5.5) implies

$$\|P_{H_\pm}Z_\pm(t)\|_{r,H_\pm} \le C\|Z_\pm(t)\|_{q,H_\pm} \le C(1+t)^{-n/2q}\|g\|_{D(A_q)}, \tag{5.16}$$

for $t \ge 0$ and $r \in (1,q]$ since $Z_\pm(t) \in L^q_{[R]}(H_\pm) \subset L^r_{[R]}(H_\pm)$ for such $r$. In view of (5.15), we deduce from (1.4) for $\Omega = H_\pm$ together with (5.16)

$$\begin{aligned}&\|\nabla v_\pm(t)\|_{q,H_\pm} \\ \le\ & Ct^{-1/2}\|a_\pm\|_{q,H_\pm} \\ &+C\|g\|_{D(A_q)}\int_0^t (t-\tau)^{-1/2}(1+t-\tau)^{-(n/r-n/q)/2}(1+\tau)^{-n/2q}d\tau \\ \le\ & Ct^{-1/2}\|g\|_q + C\|g\|_{D(A_q)}(I_1+I_2),\end{aligned}$$

for $r \in (1,q]$, where $I_1 = \int_0^{t/2}$ and $I_2 = \int_{t/2}^t$. We take $r$ so that $1 < r < \min\{n/2, q\}$. Then we see that

$$I_1 \le \left\{\begin{array}{ll} Ct^{-1/2}(1+t/2)^{-n/2r+1} & \text{if } q > n/2 \\ Ct^{-1/2}(1+t/2)^{-n/2r+1}\log(1+t/2) & \text{if } q = n/2 \\ Ct^{-1/2}(1+t/2)^{-(n/r-n/q)/2} & \text{if } q < n/2 \end{array}\right\} \le Ct^{-1/2},$$

for $t > 0$ and that

$$I_2 \le \begin{cases} C(1+t/2)^{-n/2q} & \text{if } q > n, \\ C(1+t/2)^{-1/2} & \text{if } q \le n, \end{cases}$$

for $t > 0$. Collecting the estimates above concludes (5.14). This completes the proof. □

The following lemma is concerned with the $L^\infty$ estimate of the semigroup (the restriction $q > n$ will be removed later).

**Lemma 5.5.** *Let $3 \le n < q < \infty$. There is a constant $C = C(\Omega, n, q) > 0$ such that*

$$\|e^{-tA}f\|_\infty \le Ct^{-n/2q}\|f\|_q, \tag{5.17}$$

*for $t > 0$ and $f \in L^q_\sigma(\Omega)$.*

*Proof.* For fixed $R \geq R_0 + 1$, estimate (5.4) together with the Sobolev embedding property implies

$$\|e^{-tA}f\|_{\infty,\Omega_R} \leq Ct^{-n/2q}\|f\|_q,$$

for $t \geq 2$ and $f \in L^q_\sigma(\Omega)$ on account of $n < q < \infty$. Along the lines of the proof of Lemma 5.4, one can show

$$\|e^{-tA}f\|_{\infty,\Omega_\pm\setminus\Omega_R} \leq Ct^{-n/2q}\|f\|_q, \tag{5.18}$$

for $t \geq 2$. In fact, given $f \in L^q_\sigma(\Omega)$, we take the same $g, \{u, p_\pm\}$ and $\{v_\pm, \pi_\pm\}$, and apply the $L^q$-$L^\infty$ estimate (1.3) for $\Omega = H_\pm$ to (5.15). Then, taking (5.16) into account, we get

$$\begin{aligned} &\|v_\pm(t)\|_{\infty,H_\pm} \\ \leq\ & Ct^{-n/2q}\|a_\pm\|_{q,H_\pm} \\ &+C\|g\|_{D(A_q)}\int_0^t (t-\tau)^{-n/2q}(1+t-\tau)^{-(n/r-n/q)/2}(1+\tau)^{-n/2q}d\tau, \end{aligned}$$

for $r \in (1, q]$; we now choose $r \in (1, n/2)$ to find

$$\|v_\pm(t)\|_{\infty,H_\pm} \leq Ct^{-n/2q}\|g\|_{D(A_q)},$$

for $t \geq 1$, which proves (5.18) for $t \geq 2$. We thus obtain (5.17) for $t \geq 2$. For $0 < t < 2$, we recall (5.2) to see

$$\|e^{-tA}f\|_\infty \leq C\|e^{-tA}f\|_{1,q}^{n/q}\|e^{-tA}f\|_q^{1-n/q} \leq Ct^{-n/2q}\|f\|_q.$$

The proof is complete. □

We are now in a position to prove Theorem 2.1. Abels [1] showed (1.3) for $1 < q \leq r < \infty$; when we use this result, the first step of the following proof will become shorter. However, in order to make the present paper self-contained, we do not rely on any result of [1]. We emphasize that our proof is based on (5.11) and (5.17), in other words, the other estimates follow from them.

*Proof of Theorem* 2.1. The proof is divided into four steps.

*Step* 1. First of all, we observe (1.4) for $q = r \in (1, n]$. Indeed, it follows from (5.2) for $0 < t < 2$ and (5.11) for $t \geq 2$ that

$$\|\nabla e^{-tA}f\|_q \leq Ct^{-1/2}\|f\|_q, \tag{5.19}$$

for $t > 0$ and $f \in L^q_\sigma(\Omega)$ provided $1 < q \leq n$. In this step we accomplish the proof of (1.3) for $1 < q \leq r \leq \infty$ $(q \neq \infty)$ and (1.4) for $1 < q \leq r \leq n$. We begin with the removal of the restriction $q > n$ in Lemma 5.5. In view of (5.19) and the Sobolev embedding property we have

$$\|e^{-tA}f\|_r \leq Ct^{-1/2}\|f\|_q, \tag{5.20}$$

for $t > 0$ and $f \in L^q_\sigma(\Omega)$ when $1 < q < n$ and $1/r = 1/q - 1/n$. Let $n/(k+1) < q < n/k$ with $k = 1, 2, \cdots, n-1$. We put $\{q_j\}_{j=0}^k$ in such a way that

$1/q_{j+1} = 1/q_j - 1/n$ $(j = 0, 1, \cdots, k-1)$ with $q_0 = q$. Since $n < q_k < \infty$, we make use of (5.17) with $q = q_k$ and (5.20) to obtain

$$\|e^{-tA}f\|_\infty \le Ct^{-n/2q_k}\|e^{-(t/2)A}f\|_{q_k} \le Ct^{-n/2q_k-k/2}\|f\|_q,$$

for $t > 0$, which proves (5.17) except for $q = n, n/2, \cdots, n/(n-1)$. But the exceptional cases can be also deduced via interpolation. Thus the $L^q$-$L^\infty$ estimate (5.17) has been established for all $q \in (1, \infty)$. This together with the $L^q$ boundedness (namely, (1.3) for $q = r$) immediately gives (1.3) for $1 < q \le r \le \infty$, from which combined with (5.19) we further obtain (1.4) for $1 < q \le r \le n$.

*Step* 2. In this step we prove (1.4) for $1 < q < n < r < \infty$, making use of (1.8) due to [22]. Given $r \in (n, \infty)$, we take $s \in (n/2, n)$ so that $1/s = 1/r + 1/n$. When $1 < q \le s$, an embedding relation given by Lemma 3.1 of [22] together with (1.8) implies

$$\|\nabla e^{-tA}f\|_r \le C\|\nabla^2 e^{-tA}f\|_s \le C\|Ae^{-tA}f\|_s \le Ct^{-1}\|e^{-(t/2)A}f\|_s,$$

for $t > 0$, from which together with (1.3) we obtain (1.4). If $s < q < n$, which implies $r < q_*$ with $1/q_* = 1/q - 1/n$, then by the same reasoning as above

$$\|\nabla e^{-tA}f\|_r \le \|\nabla e^{-tA}f\|_{q_*}^{1-\theta}\|\nabla e^{-tA}f\|_q^\theta \le C\|Ae^{-tA}f\|_q^{1-\theta}\|\nabla e^{-tA}f\|_q^\theta,$$

for $t > 0$, where $1/r = (1-\theta)/q_* + \theta/q = 1/q - (1-\theta)/n$. Therefore, (5.19) yields (1.4).

*Step* 3. Let $f \in L^1(\Omega) \cap L^s_\sigma(\Omega)$ for some $s \in (1, \infty)$. This step is devoted to the case $q = 1$, namely $L^1$-$L^r$ estimate. Let $1 < r < \infty$. We apply a simple duality argument; in fact, the $L^q$-$L^\infty$ estimate implies

$$|(e^{-tA}f, g)| = |(f, e^{-tA}g)| \le \|f\|_1\|e^{-tA}g\|_\infty \le Ct^{-(n-n/r)/2}\|f\|_1\|g\|_{r/(r-1)},$$

for $g \in L^{r/(r-1)}_\sigma(\Omega)$, which gives (1.3) for $q = 1 < r < \infty$. Combining this with (5.17) and (1.4), respectively, we obtain (1.3) for $q = 1 < r = \infty$ and (1.4) for $q = 1 < r < \infty$.

*Step* 4. Once the $L^q$-$L^r$ estimates (1.3) and (1.4) are established, (2.4) and (2.5) can be proved by means of a standard approximation procedure. We show only the behavior as $t \to \infty$ (which is the main concern in the present paper). Let $1 < q < \infty$ and $f \in L^q_\sigma(\Omega)$. For any $\varepsilon > 0$ we take $f_\varepsilon \in C^\infty_{0,\sigma}(\Omega)$ such that $\|f_\varepsilon - f\|_q < \varepsilon$. It then follows from (1.3) that

$$\|e^{-tA}f\|_q \le C\varepsilon + Ct^{-(n-n/q)/2}\|f_\varepsilon\|_1,$$

for $t > 0$, which immediately yields

$$\lim_{t\to\infty}\|e^{-tA}f\|_q = 0, \tag{5.21}$$

since $\varepsilon > 0$ is arbitrary (one can give another proof by use of $\ker(A_q) = \{0\}$). Let $K$ be a precompact set in $L^q_\sigma(\Omega)$. For any $\eta > 0$ there is a finite set $\{f_j\}_{j=1}^m \subset K$ so that $\{B_\eta(f_j)\}_{j=1}^m$ is a covering of $K$, where $B_\eta(f_j)$ denotes the open ball centered at $f_j$ with radius $\eta$. Then we have

$$\sup_{f\in K} \|e^{-tA}f\|_q \le C\eta + \max_{1\le j\le m} \|e^{-tA}f_j\|_q.$$

Hence, from (5.21) we deduce

$$\lim_{t\to\infty} \sup_{f\in K} \|e^{-tA}f\|_q = 0. \tag{5.22}$$

All the other decay properties as $t\to\infty$ follow from (5.22) combined with (1.3) and (1.4). We have completed the proof. □

## 6. The Navier–Stokes flow

In this section we apply the developed $L^q$-$L^r$ estimates of the semigroup to the Navier–Stokes initial value problem. In the proof of Theorems 2.2 and 2.3, we will not cite (1.3) and/or (1.4) if the application is obvious. We first prove Theorem 2.2.

*Proof of Theorem* 2.2. One can construct a unique global solution $u(t)$ of the integral equation

$$u(t) = e^{-tA}a - \int_0^t e^{-(t-\tau)A}P(u\cdot\nabla u)(\tau)d\tau, \quad t>0, \tag{6.1}$$

by means of a standard contraction mapping principle, in exactly the same way as in Kato [39], provided that $\|a\|_n \le \delta_0$, where $\delta_0 = \delta_0(\Omega, n) > 0$ is a constant. The solution $u(t)$ satisfies

$$\|u(t)\|_r \le Ct^{-1/2+n/2r}\|a\|_n \quad \text{for } n\le r\le\infty, \tag{6.2}$$

$$\|\nabla u(t)\|_n \le Ct^{-1/2}\|a\|_n, \tag{6.3}$$

for $t>0$ together with the singular behavior

$$\|u(t)\|_r = o\left(t^{-1/2+n/2r}\right) \quad \text{for } n<r\le\infty;\ \|\nabla u(t)\|_n = o\left(t^{-1/2}\right), \tag{6.4}$$

as $t\to 0$. Furthermore, due to the Hölder estimate (6.9) below which is implied by (6.2) and (6.3), the solution $u(t)$ becomes actually a strong one of (1.1) with (2.6) (see [26], [32] and [64]). We now prove

$$\lim_{t\to\infty} \|u(t)\|_n = 0, \tag{6.5}$$

for still smaller $a\in L^n_\sigma(\Omega)$. To this end, we derive a certain decay property of $u(t)$, which is weaker than (2.11) but sufficient for the proof of (6.5), assuming additionally $a\in L^1(\Omega)\cap L^n_\sigma(\Omega)$ with small $\|a\|_n$. Given $\gamma\in(0,1/2)$, we take $q\in(n/2,n)$ so that $\gamma = n/2q - 1/2$; then,

$$\|u(t)\|_n \le Ct^{-\gamma}\|a\|_q + C\int_0^t (t-\tau)^{-1/2}\|u(\tau)\|_n\|\nabla u(\tau)\|_n d\tau,$$

which together with (6.3) implies

$$t^\gamma\|u(t)\|_n \le C\|a\|_q + C\|a\|_n \sup_{0<\tau\le t} \tau^\gamma\|u(\tau)\|_n,$$

for $t > 0$. Hence, for any $\gamma \in (0, 1/2)$ there are constants $\delta_* = \delta_*(\Omega, n, \gamma) \in (0, \delta_0]$ and $C = C(\Omega, n, \gamma) > 0$ such that if $\|a\|_n \le \delta_*$, then $\|u(t)\|_n \le Ct^{-\gamma}\|a\|_q$ for $t > 0$, which together with (6.2) yields

$$\|u(t)\|_n \le C(1+t)^{-\gamma}(\|a\|_1 + \|a\|_n), \tag{6.6}$$

for $t \ge 0$ (this decay rate is not sharp and will be improved in Theorem 2.3). From now on we fix $\gamma \in (0, 1/2)$ and set $\delta = \delta_*(\Omega, n, \gamma)/2$. Given $a \in L^n_\sigma(\Omega)$ with $\|a\|_n \le \delta$ and any $\varepsilon \in (0, \delta]$, we take $a_\varepsilon \in C^\infty_{0,\sigma}(\Omega)$ so that $\|a_\varepsilon - a\|_n < \varepsilon$. Since $\|a_\varepsilon\|_n \le \delta_*$, the corresponding global solution fulfills (6.6). We combine this fact with the continuous dependence: $L^n_\sigma(\Omega) \ni u(0) \mapsto u \in BC([0, \infty); L^n_\sigma(\Omega))$, where $BC$ denotes the class of bounded continuous functions. As a consequence, the global solution $u(t)$ with $u(0) = a$ satisfies $\|u(t)\|_n \le C\varepsilon + C(1+t)^{-\gamma}$, which proves (6.5) (although the method above was mentioned in [39] and is well known, we gave the proof for completeness; see also Theorem 3 of Wiegner [66] for another proof). Combining (6.5) with (6.2) for $r = \infty$ immediately leads us to (2.8) for $n \le r < \infty$. We next prove (2.8) for $r = \infty$ and (2.9). As is standard, we rewrite the integral equation (6.1) in the form

$$u(t) = e^{-(t/2)A}u(t/2) - \int_{t/2}^t e^{-(t-\tau)A}P(u\cdot\nabla u)(\tau)d\tau, \quad t > 0. \tag{6.7}$$

Then we obtain

$$\begin{aligned} &\|u(t)\|_\infty + \|\nabla u(t)\|_n \\ \le\;& Ct^{-1/2}\|u(t/2)\|_n + C\int_{t/2}^t (t-\tau)^{-3/4}\|u(\tau)\|_{2n}\|\nabla u(t)\|_n d\tau, \end{aligned}$$

from which together with (6.3) we at once deduce

$$t^{1/2}(\|u(t)\|_\infty + \|\nabla u(t)\|_n) \le C\|u(t/2)\|_n + C\|a\|_n \sup_{t/2\le\tau\le t} \tau^{1/4}\|u(\tau)\|_{2n},$$

for $t > 0$. Obviously, (6.5) and (2.8) for $r = 2n$ conclude both (2.8) for $r = \infty$ and (2.9). These immediately yield

$$\|P(u\cdot\nabla u)(t)\|_n \le C\|u(t)\|_\infty\|\nabla u(t)\|_n = o(t^{-1}), \tag{6.8}$$

as $t \to \infty$, which will be used to show (2.10) below. Fix $\theta \in (0, 1/2)$ arbitrarily. We then observe

$$\|u(t) - u(\tau)\|_\infty + \|\nabla u(t) - \nabla u(\tau)\|_n \le C(t-\tau)^\theta\tau^{-1/2-\theta}\|a\|_n, \tag{6.9}$$

for $0 < \tau < t$. Given $f \in L^q_\sigma(\Omega), 1 < q \le n$, we have

$$\begin{aligned}
&\|e^{-tA}f - e^{-\tau A}f\|_\infty + \|\nabla e^{-tA}f - \nabla e^{-\tau A}f\|_n \\
\le\ & \int_\tau^t \left(\|e^{-(s/2)A}Ae^{-(s/2)A}f\|_\infty + \|\nabla e^{-(s/2)A}Ae^{-(s/2)A}f\|_n\right) ds \\
\le\ & C\int_\tau^t s^{-n/2q}\|Ae^{-(s/2)A}f\|_q ds \\
\le\ & C(t-\tau)\,\tau^{-n/2q-1}\|f\|_q,
\end{aligned}$$

which implies

$$\|e^{-tA}f - e^{-\tau A}f\|_\infty + \|\nabla e^{-tA}f - \nabla e^{-\tau A}f\|_n \le C(t-\tau)^\theta\tau^{-n/2q-\theta}\|f\|_q,$$

for $0 < \tau < t$, where $0 < \theta < 1$. Using this estimate together with

$$\begin{aligned}
u(t) - u(\tau) =\ & (e^{-tA} - e^{-\tau A})a - \int_\tau^t e^{-(t-s)A}P(u\cdot\nabla u)(s)ds \\
& -\int_0^\tau \{e^{-(t-s)A} - e^{-(\tau-s)A}\}P(u\cdot\nabla u)(s)ds,
\end{aligned}$$

we obtain

$$\begin{aligned}
&\|u(t) - u(\tau)\|_\infty + \|\nabla u(t) - \nabla u(\tau)\|_n \\
\le\ & C(t-\tau)^\theta\tau^{-1/2-\theta}\|a\|_n + C\int_\tau^t (t-s)^{-n/2p-1/2}\|u(s)\|_p\|\nabla u(s)\|_n ds \\
& +C\int_0^\tau (t-\tau)^\theta(\tau-s)^{-n/2r-1/2-\theta}\|u(s)\|_r\|\nabla u(s)\|_n ds,
\end{aligned}$$

where we have taken $p, r \in (n,\infty)$ so that $0 < n/2r < n/2p = 1/2 - \theta$. Then (6.2) and (6.3) yield (6.9). From (6.7) and (6.9) one can deduce the representation

$$Au(t) = Ae^{-(t/2)A}u(t/2) + \{e^{-(t/2)A} - 1\}P(u\cdot\nabla u)(t) + Az(t), \tag{6.10}$$

in $L^n_\sigma(\Omega)$, where

$$z(t) = \int_{t/2}^t e^{-(t-\tau)A}P\{(u\cdot\nabla u)(t) - (u\cdot\nabla u)(\tau)\}d\tau.$$

In fact, (6.9) implies

$$\begin{aligned}
\|Az(t)\|_n &\le C\int_{t/2}^t (t-\tau)^{-1+\theta}\tau^{-1/2-\theta}(\|\nabla u(t)\|_n + \|u(\tau)\|_\infty)d\tau \\
&\le Ct^{-1/2}\|\nabla u(t)\|_n + Ct^{-1}\sup_{t/2\le\tau\le t}\tau^{1/2}\|u(\tau)\|_\infty,
\end{aligned}$$

for $t > 0$. As a direct consequence of (2.8) for $r = \infty$ and (2.9), we see that $\|Az(t)\|_n = o(t^{-1})$, as $t \to \infty$. In view of (6.10), we collect (6.5), (6.8) and the above decay property of $Az(t)$ to obtain $\|Au(t)\|_n = o(t^{-1})$ as $t \to \infty$, which together with (6.8) again shows (2.10). The proof is complete. □

**Remark 6.1.** Consider briefly the 3-dimensional stability problem mentioned in Remark 2.5. The problem is reduced to the global existence and asymptotic behavior of the solution to

$$u(t) = e^{-tA}a - \int_0^t e^{-(t-\tau)A}P(u\cdot\nabla u + w\cdot\nabla u + u\cdot\nabla w)(\tau)d\tau, \quad t>0,$$

where $w$ is a stationary solution of class $\nabla w \in L^r(\Omega), 1<r\le 2$, and $a\in L^3_\sigma(\Omega)$ is a given initial disturbance. Set

$$E(t) = \sup_{0<\tau\le t} \tau^{1/2}(\|u(\tau)\|_\infty + \|\nabla u(\tau)\|_3) + \sup_{0<\tau\le t}\tau^{1/4}\|u(\tau)\|_6,$$

and fix $r\in(1,3/2)$ arbitrarily. Then the integral equation yields the a priori estimate

$$E(t)\le C\|a\|_3 + CE(t)^2 + C(\|\nabla w\|_r + \|\nabla w\|_2)E(t),$$

for $t>0$, which gives an affirmative answer to the stability problem provided that both $\|\nabla w\|_r + \|\nabla w\|_2$ and $\|a\|_3$ are small enough. In fact, by following the argument of Chen [13], the above inequality for $E(t)$ is deduced from

$$\begin{aligned}&\|e^{-tA}P(w\cdot\nabla u + u\cdot\nabla w)\|_\infty + \|\nabla e^{-tA}P(w\cdot\nabla u+u\cdot\nabla w)\|_3\\ \le\;& C(\|\nabla w\|_r + \|\nabla w\|_2)(\|u\|_\infty + \|\nabla u\|_3)\, t^{-3/4}(1+t)^{-3/2r+3/4},\end{aligned}$$

and

$$\begin{aligned}&\|e^{-tA}P(w\cdot\nabla u+u\cdot\nabla w)\|_6\\ \le\;& C(\|\nabla w\|_r + \|\nabla w\|_2)(\|u\|_\infty + \|\nabla u\|_3)\, t^{-1/2}(1+t)^{-3/2r+3/4}.\end{aligned}$$

We next assume that $a\in L^1(\Omega)\cap L^n_\sigma(\Omega)$ with $\|a\|_n\le\delta$. Let $u(t)$ be the global solution constructed in Theorem 2.2. Our particular concern is more rapid decay properties of $u(t)$. Starting from (6.2) and (6.3), we observe that $u(t)\in W^{1,r}(\Omega)$ for $1<r<n$ and $t>0$ (without going back to approximate solutions). In fact, there is a constant $M = M(\Omega,n,r,\|a\|_1,\|a\|_n,T)>0$ such that

$$\begin{aligned}\|\nabla^j u(t)\|_r &\le Ct^{-j/2}\|a\|_r + C\int_0^t (t-\tau)^{-1+n/2r-j/2}\|u(\tau)\|_n\|\nabla u(\tau)\|_n d\tau\\ &\le Mt^{-j/2},\end{aligned}$$

for $n/2\le r<n$, $j=0,1$ and $0<t\le T$, where $T>0$ is arbitrarily fixed; and then,

$$\begin{aligned}\|\nabla^j u(t)\|_r &\le Ct^{-j/2}\|a\|_r + C\int_0^t (t-\tau)^{-j/2}\|u(\tau)\|_{2r}\|\nabla u(\tau)\|_{2r} d\tau\\ &\le Mt^{-j/2},\end{aligned}$$

for $n/4\le r<n/2$, $j=0,1$ and $0<t\le T$. We repeat the process above to get $u(t)\in W^{1,r}(\Omega)$ for $1<r<n$ with

$$\sup_{0<t\le T}(\|u(t)\|_r + t^{1/2}\|\nabla u(t)\|_r)\le M. \tag{6.11}$$

**Remark 6.2.** Following the argument of Kato [39], we see that the above constant $M$ does not depend on $T>0$ if $\|a\|_n$ is still smaller. However, we do not rely on his procedure because the smallness of initial data depends on $r > 1$. Note that the constant $\eta$ in Theorem 2.3 is independent of $r > 1$.

As the first step of our proof of Theorem 2.3, we show the following lemma which gives a little slower decay rate than desired (later on, $\varepsilon > 0$ will be removed so that estimates will become sharp).

**Lemma 6.1.** *Let $n \geq 3$ and $a \in L^1(\Omega) \cap L^n_\sigma(\Omega)$. For any small $\varepsilon > 0$ there are constants $\eta_* = \eta_*(\Omega, n, \varepsilon) \in (0, \delta]$ and $C = C(\Omega, n, \|a\|_1, \|a\|_n, \varepsilon) > 0$ such that if $\|a\|_n \leq \eta_*$, then the solution $u(t)$ obtained in Theorem 2.2 satisfies*

$$\|u(t)\|_{n/(n-1)} \leq C(1+t)^{-1/2+\varepsilon}, \tag{6.12}$$

$$\|u(t)\|_{2n} \leq Ct^{-1/4}(1+t)^{-n/2+1/2+\varepsilon}, \tag{6.13}$$

$$\|\nabla u(t)\|_n \leq Ct^{-1/2}(1+t)^{-n/2+1/2+\varepsilon}, \tag{6.14}$$

*for $t > 0$.*

*Proof.* We make use of (1.3) for $r = \infty$ to obtain

$$\begin{aligned} &|(e^{-(t-\tau)A}P(u\cdot\nabla u)(\tau), \varphi)| = |((u\cdot\nabla u)(\tau), e^{-(t-\tau)A}\varphi)| \\ \leq\ & C(t-\tau)^{-(n-n/q)/2}\|u(\tau)\|_{n/(n-1)}\|\nabla u(\tau)\|_n\|\varphi\|_{q/(q-1)}, \end{aligned}$$

for all $\varphi \in C^\infty_{0,\sigma}(\Omega)$, which gives

$$\begin{aligned} &\|\nabla^j e^{-(t-\tau)A}P(u\cdot\nabla u)(\tau)\|_q \\ \leq\ & C(t-\tau)^{-(n-n/q)/2-j/2}\|u(\tau)\|_{n/(n-1)}\|\nabla u(\tau)\|_n, \end{aligned} \tag{6.15}$$

for $1 < q < \infty$, $j = 0, 1$ and $0 < \tau < t$ (the case $j = 1$ follows from (1.4) and the case $j = 0$). Given $\varepsilon > 0$, we take $p \in (1, n/(n-1))$ so that $1/p = 1 - 2\varepsilon/n$. From (6.15) with $q = n/(n-1)$ it follows that

$$\|u(t)\|_{n/(n-1)} \leq Ct^{-1/2+\varepsilon}\|a\|_p + C\int_0^t (t-\tau)^{-1/2}\|u(\tau)\|_{n/(n-1)}\|\nabla u(\tau)\|_n d\tau.$$

In an analogous way to the deduction of (6.6), one can take a constant $\eta_0 = \eta_0(\Omega, n, \varepsilon) \in (0, \delta]$ such that if $\|a\|_n \leq \eta_0$, then $\|u(t)\|_{n/(n-1)} \leq Ct^{-1/2+\varepsilon}\|a\|_p$ for $t > 0$, which together with (6.11) gives (6.12). To show (6.13) and (6.14), we will derive

$$\|\nabla u(t)\|_r \leq Ct^{-(n-n/r)/2-1/2+\varepsilon} \quad \text{for } r = n,\ 2n/3, \tag{6.16}$$

for $t > 0$. We divide the integral of (6.1) into two parts

$$\int_0^t e^{-(t-\tau)A}P(u\cdot\nabla u)(\tau)d\tau = \int_0^{t/2} + \int_{t/2}^t = v(t) + w(t); \tag{6.17}$$

then we obtain

$$\|\nabla u(t)\|_r \leq Ct^{-(n-n/r)/2-1/2+\varepsilon}\|a\|_p + I_1 + I_2,$$

for $t > 0$ ($p$ is the same as above) with

$$I_1 = \|\nabla v(t)\|_r \leq C\int_0^{t/2}(t-\tau)^{-(n-n/r)/2-1/2}\|u(\tau)\|_{n/(n-1)}\|\nabla u(\tau)\|_n d\tau,$$

$$I_2 = \|\nabla w(t)\|_r \le C \int_{t/2}^{t} (t-\tau)^{-1/2}\|u(\tau)\|_\infty \|\nabla u(\tau)\|_r d\tau,$$

where (6.15) has been used in $I_1$. Using (6.12) together with (6.2) and (6.3), we see that

$$I_1 \le Ct^{-(n-n/r)/2-1/2+\varepsilon}\|a\|_n, \quad I_2 \le C\|a\|_n \sup_{t/2\le\tau\le t} \|\nabla u(t)\|_r,$$

for $t > 0$. Therefore, setting

$$E_r(t) = \sup_{0<\tau\le t} \tau^{(n-n/r)/2+1/2-\varepsilon}\|\nabla u(\tau)\|_r \quad \text{for } r = n,\ 2n/3,$$

we get $E_r(t) \le C\|a\|_p + C\|a\|_n + C_0\|a\|_n E_r(t)$ for $t > 0$, where $C_0 > 0$ is independent of $a$. As a consequence, there is a constant $\eta_* = \eta_*(\Omega, n, \varepsilon) \in (0, \eta_0]$ such that if $\|a\|_n \le \eta_*$, then $E_n(t) + E_{2n/3}(t) \le C$ for $t > 0$, which proves (6.16). This combined with the Sobolev embedding, (6.2) for $r = 2n$ and (6.3) imply (6.13) and (6.14). □

Based on Lemma 6.1, we supply the proof of Theorem 2.3, by which we conclude this paper.

*Proof of Theorem* 2.3. We fix $\varepsilon \in (0, 1/2)$ and put $\eta = \eta(\Omega, n) = \eta_*(\Omega, n, \varepsilon)$. Assuming $\|a\|_n \le \eta$, we first show (2.11). Since

$$\|e^{-tA}a\|_r \le Ct^{-(n-n/r)/2}\|a\|_1,$$

for $t > 0$, our task is to derive the required estimate of (6.17). By (6.15) together with (6.12) and (6.14) we have

$$\begin{aligned}\|v(t)\|_r &\le C \int_0^{t/2} (t-\tau)^{-(n-n/r)/2}\|u(\tau)\|_{n/(n-1)}\|\nabla u(\tau)\|_n d\tau \\ &\le Ct^{-(n-n/r)/2} \int_0^\infty \tau^{-1/2}(1+\tau)^{-n/2+2\varepsilon} d\tau \\ &\le Ct^{-(n-n/r)/2},\end{aligned}$$

for $1 < r \le \infty$ and $t > 0$; here, note that the case $r = \infty$ follows from the $L^q$-$L^\infty$ estimate (1.3) together with (6.15). If $1 < r < n/(n-2)$, then the same estimate of the integrand as above works well on $w(t)$ too; as a result, we have

$$\|w(t)\|_r \le Ct^{-(n-n/r)/2-n/2+1/2+2\varepsilon},$$

for $t > 0$. For $r = \infty$, we make use of (6.13) and (6.14) to get

$$\|w(t)\|_\infty \le C \int_{t/2}^{t} (t-\tau)^{-3/4}\|u(\tau)\|_{2n}\|\nabla u(\tau)\|_n d\tau \le Ct^{-n+1/2+2\varepsilon},$$

for $t > 0$. We collect the estimates above to obtain (2.11) for $1 < r < n/(n-2)$ and $r = \infty$; and the remaining case $n/(n-2) \le r < \infty$ follows via interpolation as well.

We next show (2.12). Let $1 < r \le n$. In view of (6.7), we have

$$\|\nabla u(t)\|_r \le Ct^{-1/2}\|u(t/2)\|_r + \|\nabla w(t)\|_r,$$

for $t > 0$, where $w(t)$ is the same as above. By (2.11) the proof is reduced to the estimate of $\|\nabla w(t)\|_r$. If in particular $1 < r < n/(n-1)$, then from (2.11), (6.14) and (6.15) we deduce

$$\begin{aligned}\|\nabla w(t)\|_r &\le C\int_{t/2}^t (t-\tau)^{-(n-n/r)/2-1/2}\|u(\tau)\|_{n/(n-1)}\|\nabla u(\tau)\|_n d\tau\\ &\le Ct^{-(n-n/r)/2-n/2+\varepsilon},\end{aligned}$$

for $t > 0$. If $r = n$, then one appeals again to (2.11) and (6.14) to find

$$\|\nabla w(t)\|_n \le C\int_{t/2}^t (t-\tau)^{-1/2}\|u(\tau)\|_\infty\|\nabla u(\tau)\|_n d\tau \le Ct^{-n+1/2+\varepsilon},$$

for $t > 0$. We thus obtain (2.12) for $1 < r < n/(n-1)$ and $r = n$; and the case $n/(n-1) \le r < n$ also follows via interpolation. It remains to show the case $n < r < \infty$. From (1.4) for $1 < q < n < r < \infty$ we deduce

$$\|\nabla u(t)\|_r \le Ct^{-(n/q-n/r)/2-1/2}\|u(t/2)\|_q + \|\nabla w(t)\|_r,$$

for $t > 0$, and the first term possesses the desired decay property on account of (2.11). We take $p$ in such a way that $1/n < 1/p < 1/n + 1/r$. Since we have already known (2.12) for $r = p$ as well as (2.11), we are led to

$$\begin{aligned}\|\nabla w(t)\|_r &\le C\int_{t/2}^t (t-\tau)^{-(n/p-n/r)/2-1/2}\|u(\tau)\|_\infty\|\nabla u(\tau)\|_p d\tau\\ &\le Ct^{-n+n/2r},\end{aligned}$$

for $t > 0$, which proves (2.12) for $n < r < \infty$.

Finally, by use of (6.10), we show (2.13) and thereby (2.14), (2.15) and (2.16). From (2.11) and (2.12) it follows that

$$\|Ae^{-(t/2)A}u(t/2)\|_r \le Ct^{-1}\|u(t/2)\|_r = O\left(t^{-(n-n/r)/2-1}\right), \tag{6.18}$$

as $t \to \infty$ and that

$$\|P(u\cdot\nabla u)(t)\|_r \le C\|u(t)\|_\infty\|\nabla u(t)\|_r = O\left(t^{-n+n/2r-1/2}\right), \tag{6.19}$$

as $t \to \infty$. We are thus going to estimate

$$\begin{aligned}\|Az(t)\|_r &\le C\|\nabla u(t)\|_r\int_{t/2}^t (t-\tau)^{-1}\|u(t)-u(\tau)\|_\infty d\tau\\ &\quad + C\int_{t/2}^t (t-\tau)^{-1}\|u(\tau)\|_\infty\|\nabla u(t)-\nabla u(\tau)\|_r d\tau = I_1 + I_2.\end{aligned}$$

With the aid of (6.9) and (2.12) we observe

$$I_1 = O\left(t^{-(n-n/r)/2-1}\right), \tag{6.20}$$

as $t \to \infty$. We need also a Hölder estimate of $\nabla u(t)$ in $L^r(\Omega)$, which implies the decay property of $I_2$ as well as $u \in C(0,\infty; D(A_r)) \cap C^1(0,\infty; L^r_\sigma(\Omega))$. To this end, let us consider

$$\begin{aligned}
& u(t) - u(\tau) \\
= \; & \{e^{-(t-\tau/2)A} - e^{-(\tau/2)A}\}u(\tau/2) - \int_\tau^t e^{-(t-s)A}P(u\cdot\nabla u)(s)ds \\
& - \int_{\tau/2}^\tau \{e^{-(t-s)A} - e^{-(\tau-s)A}\}P(u\cdot\nabla u)(s)ds \\
= \; & w_1(t,\tau) + w_2(t,\tau) + w_3(t,\tau),
\end{aligned}$$

for $0<\tau<t$. In order to estimate $w_1$, we take $q\in(1,n)$ (resp. $q\in(1,r]$) if $r>n$ (resp. $r\le n$); then, given $f\in L^q_\sigma(\Omega)$, we have

$$\begin{aligned}
& \|\nabla\{e^{-(t-\tau/2)A} - e^{-(\tau/2)A}\}f\|_r \\
\le \; & \int_{\tau/2}^{t-\tau/2} \|\nabla e^{-(s/2)A}Ae^{-(s/2)A}f\|_r ds \\
\le \; & C\int_{\tau/2}^{t-\tau/2} s^{-(n/q-n/r)/2-1/2}\|Ae^{-(s/2)A}f\|_q ds \\
\le \; & C(t-\tau)\,\tau^{-(n/q-n/r)/2-3/2}\|f\|_q,
\end{aligned}$$

which implies

$$\|\nabla\{e^{-(t-\tau/2)A} - e^{-(\tau/2)A}\}f\|_r \le C(t-\tau)^\theta\tau^{-(n/q-n/r)/2-1/2-\theta}\|f\|_q,$$

where $0<\theta<1$. Setting $f=u(\tau/2)$ and using (2.11), we get

$$\|\nabla w_1(t,\tau)\|_r \le C(t-\tau)^\theta\tau^{-(n-n/r)/2-1/2-\theta}. \tag{6.21}$$

In order to estimate $w_2$ and $w_3$, we take $q\in(1,r]$ so that $0<1/q-1/n<1/r$; then we see from (6.19) in $L^q_\sigma(\Omega)$ that

$$\|\nabla w_2(t,\tau)\|_r \le C(t-\tau)^{1/2-(n/q-n/r)/2}\tau^{-n+n/2q-1/2}, \tag{6.22}$$

and that

$$\begin{aligned}
& \|\nabla w_3(t,\tau)\|_r \\
\le \; & C\int_{\tau/2}^\tau (t-\tau)^\theta(\tau-s)^{-(n/q-n/r)/2-1/2-\theta}\|P(u\cdot\nabla u)(s)\|_q ds \\
\le \; & C(t-\tau)^\theta\tau^{-n+n/2r-\theta},
\end{aligned} \tag{6.23}$$

where $0<\theta<1/2-(n/q-n/r)/2$. Collecting (6.21), (6.22) and (6.23) together with (2.11) yields

$$I_2 = O\left(t^{-n+n/2r-1/2}\right), \tag{6.24}$$

as $t\to\infty$. From (6.18), (6.19), (6.20) and (6.24) we obtain (2.13). Due to (1.8) and in view of the equation (1.1), we deduce (2.14) immediately from (2.11), (2.12) and (2.13). By Lemma 3.1 of [22] there exist $p_\pm(t)\in\boldsymbol{R}$ such that $\|p(t)-p_\pm(t)\|_{r,\Omega_\pm} + |p_+(t)-p_-(t)| \le C\|\nabla p(t)\|_q$ for $1<q<n$ and $1/r=1/q-1/n$. Hence, (2.14) implies (2.15) and (2.16). The proof is complete. □

**Acknowledgments**

The present work was done while I stayed at Technische Universität Darmstadt as a research fellow of the Alexander von Humboldt Foundation. Their kind support and hospitality are gratefully acknowledged. I wish to express my sincere gratitude to Professor R. Farwig for his interest in this study and useful comments. I am grateful to Professor Y. Shibata who introduced me to the method in [40]. Thanks are also due to Drs. M. Franzke and H. Abels for showing me their manuscripts prior to publications.

## References

[1] H. Abels, $L_q$-$L_r$ estimates for the non-stationary Stokes equations in an aperture domain, *Z. Anal. Anwendungen* **21** (2002), 159–178.

[2] M.E. Bogovskiĭ, Solution of the first boundary value problem for the equation of continuity of an incompressible medium, *Dokl. Akad. Nauk SSSR* **248** (1979), 1037–1040; English Transl.: *Soviet Math. Dokl.* **20** (1979), 1094–1098.

[3] W. Borchers, G.P. Galdi and K. Pileckas, On the uniqueness of Leray–Hopf solutions for the flow through an aperture, *Arch. Rational Mech. Anal.* **122** (1993), 19–33.

[4] W. Borchers and T. Miyakawa, $L^2$ decay for the Navier–Stokes flow in halfspaces, *Math. Ann.* **282** (1988), 139–155.

[5] W. Borchers and T. Miyakawa, Algebraic $L^2$ decay for Navier–Stokes flows in exterior domains, *Acta Math.* **165** (1990), 189–227.

[6] W. Borchers and T. Miyakawa, On some coercive estimates for the Stokes problem in unbounded domains, *The Navier–Stokes Equations II – Theory and Numerical Methods*, 71–84, *Lecture Notes in Math.* **1530**, Springer, Berlin, 1992.

[7] W. Borchers and T. Miyakawa, On stability of exterior stationary Navier–Stokes flows, *Acta Math.* **174** (1995), 311–382.

[8] W. Borchers and K. Pileckas, Existence, uniqueness and asymptotics of steady jets, *Arch. Rational Mech. Anal.* **120** (1992), 1–49.

[9] W. Borchers and H. Sohr, On the semigroup of the Stokes operator for exterior domains in $L^q$-spaces, *Math. Z.* **196** (1987), 415–425.

[10] W. Borchers and H. Sohr, On the equations rot $v = g$ and div $u = f$ with zero boundary conditions, *Hokkaido Math. J.* **19** (1990), 67–87.

[11] W. Borchers and W. Varnhorn, On the boundedness of the Stokes semigroup in two-dimensional exterior domains, *Math. Z.* **213** (1993), 275–299.

[12] H. Brezis, Remarks on the preceding paper by M. Ben-Artzi "Global solutions of two-dimensional Navier–Stokes and Euler equations", *Arch. Rational Mech. Anal.* **128** (1994), 359–360.

[13] Z.M. Chen, Solutions of the stationary and nonstationary Navier–Stokes equations in exterior domains, *Pacific J. Math.* **159** (1993), 227–240.

[14] W. Dan, T. Kobayashi and Y. Shibata, On the local energy decay approach to some fluid flow in an exterior domain, *Recent Topics on Mathematical Theory of Viscous Incompressible Fluid*, 1–51, *Lecture Notes Numer. Appl. Math.* **16**, Kinokuniya, Tokyo, 1998.

[15] W. Dan and Y. Shibata, On the $L_q$-$L_r$ estimates of the Stokes semigroup in a 2-dimensional exterior domain, *J. Math. Soc. Japan* **51** (1999), 181–207.

[16] W. Dan and Y. Shibata, Remarks on the $L_q$-$L_\infty$ estimate of the Stokes semigroup in a 2-dimensional exterior domain, *Pacific J. Math.* **189** (1999), 223–239.

[17] W. Desch, M. Hieber and J. Prüss, $L^p$ theory of the Stokes equation in a half space, *J. Evol. Equations* **1** (2001), 115–142.

[18] R. Farwig, Note on the flux condition and pressure drop in the resolvent problem of the Stokes system, *Manuscripta Math.* **89** (1996), 139–158.

[19] R. Farwig and H. Sohr, An approach to resolvent estimates for the Stokes equations in $L^q$-spaces, *The Navier–Stokes Equations II – Theory and Numerical Methods*, 97–110, *Lecture Notes in Math.* **1530**, Springer, Berlin, 1992.

[20] R. Farwig and H. Sohr, Generalized resolvent estimates for the Stokes system in bounded and unbounded domains, *J. Math. Soc. Japan* **46** (1994), 607–643.

[21] R. Farwig and H. Sohr, On the Stokes and Navier–Stokes system for domains with noncompact boundary in $L^q$-spaces, *Math. Nachr.* **170** (1994), 53–77.

[22] R. Farwig and H. Sohr, Helmholtz decomposition and Stokes resolvent system for aperture domains in $L^q$-spaces, *Analysis* **16** (1996), 1–26.

[23] M. Franzke, Strong solutions of the Navier–Stokes equations in aperture domains, *Ann. Univ. Ferrara Sez. VII. Sc. Mat.* **46** (2000), 161–173.

[24] M. Franzke, *Die Navier–Stokes-Gleichungen in Öffnungsgebieten*, Doktorschrift, Technische Universität Darmstadt, 2000.

[25] M. Franzke, Strong $L^q$-theory of the Navier–Stokes equations in aperture domains, Preprint Nr. 2139 TU Darmstadt (2001).

[26] H. Fujita and T. Kato, On the Navier–Stokes initial value problem. I, *Arch. Rational Mech. Anal.* **16** (1964), 269–315.

[27] D. Fujiwara and H. Morimoto, An $L_r$-theorem of the Helmholtz decomposition of vector fields, *J. Fac. Sci. Univ. Tokyo Sect. IA* **24** (1977), 685–700.

[28] G.P. Galdi, *An Introduction to the Mathematical Theory of the Navier–Stokes Equations, Vol. I: Linearized Steady Problems, Vol. II: Nonlinear Steady Problems*, Springer, New York, 1994.

[29] Y. Giga, Analyticity of the semigroup generated by the Stokes operator in $L_r$ spaces, *Math. Z.* **178** (1981), 297–329.

[30] Y. Giga, Domains of fractional powers of the Stokes operator in $L_r$-spaces, *Arch. Rational Mech. Anal.* **89** (1985), 251–265.

[31] Y. Giga, S. Matsui and Y. Shimizu, On estimates in Hardy spaces for the Stokes flow in a half space, *Math. Z.* **231** (1999), 383–396.

[32] Y. Giga and T. Miyakawa, Solutions in $L_r$ of the Navier-Stokes initial value problem, *Arch. Rational Mech. Anal.* **89** (1985), 267–281.

[33] Y. Giga and H. Sohr, On the Stokes operator in exterior domains, *J. Fac. Sci. Univ. Tokyo Sect. IA* **36** (1989), 103–130.

[34] J.G. Heywood, On uniqueness questions in the theory of viscous flow, *Acta Math.* **136** (1976), 61–102.

[35] J.G. Heywood, Auxiliary flux and pressure conditions for Navier–Stokes problems, *Approximation Methods for Navier–Stokes Problems*, 223–234, *Lecture Notes in Math.* **771**, Springer, Berlin, 1980.

[36] J.G. Heywood, The Navier–Stokes equations: on the existence, regularity and decay of solutions, *Indiana Univ. Math. J.* **29** (1980), 639–681.

[37] T. Hishida, On a class of stable steady flows to the exterior convection problem, *J. Differential Equations* **141** (1997), 54–85.

[38] H. Iwashita, $L_q$-$L_r$ estimates for solutions of the nonstationary Stokes equations in an exterior domain and the Navier–Stokes initial value problems in $L_q$ spaces, *Math. Ann.* **285** (1989), 265–288.

[39] T. Kato, Strong $L^p$ solutions of the Navier–Stokes equation in $\boldsymbol{R}^m$, with applications to weak solutions, *Math. Z.* **187** (1984), 471–480.

[40] T. Kobayashi and Y. Shibata, On the Oseen equation in the three dimensional exterior domains, *Math. Ann.* **310** (1998), 1–45.

[41] H. Kozono, Global $L^n$-solution and its decay property for the Navier–Stokes equations in half-space $\boldsymbol{R}^n_+$, *J. Differential Equations* **79** (1989), 79–88.

[42] H. Kozono, $L^1$-solutions of the Navier–Stokes equations in exterior domains, *Math. Ann.* **312** (1998), 319–340.

[43] H. Kozono, Rapid time-decay and net force to the obstacles by the Stokes flow in exterior domains, *Math. Ann.* **320** (2001), 709–730.

[44] H. Kozono and T. Ogawa, Two-dimensional Navier–Stokes flow in unbounded domains, *Math. Ann.* **297** (1993), 1–31.

[45] H. Kozono and T. Ogawa, Global strong solution and its decay properties for the Navier–Stokes equations in three dimensional domains with non-compact boundaries, *Math. Z.* **216** (1994), 1–30.

[46] H. Kozono and T. Ogawa, On stability of Navier–Stokes flows in exterior domains, *Arch. Rational Mech Anal.* **128** (1994), 1–31.

[47] H. Kozono and H. Sohr, Global strong solution of the Navier–Stokes equations in 4 and 5 dimensional unbounded domains, *Ann. Inst. H. Poincaré Anal. Nonlinéaire* **16** (1999), 535–561.

[48] H. Kozono and M. Yamazaki, Local and global unique solvability of the Navier–Stokes exterior problem with Cauchy data in the space $L^{n,\infty}$, *Houston J. Math.* **21** (1995), 755–799.

[49] H. Kozono and M. Yamazaki, On a larger class of stable solutions to the Navier–Stokes equations in exterior domains, *Math. Z.* **228** (1998), 751–785.

[50] P. Maremonti and V.A. Solonnikov, On nonstationary Stokes problem in exterior domains, *Ann. Sc. Norm. Sup. Pisa* **24** (1997), 395–449.

[51] K. Masuda, Weak solutions of Navier–Stokes equations, *Tôhoku Math. J.* **36** (1984), 623–646.

[52] M. McCracken, The resolvent problem for the Stokes equation on halfspaces in $L_p$, *SIAM J. Math. Anal.* **12** (1981), 201–228.

[53] T. Miyakawa, The Helmholtz decomposition of vector fields in some unbounded domains, *Math. J. Toyama Univ.* **17** (1994), 115–149.

[54] K. Pileckas, Three-dimensional solenoidal vectors, *Zap. Nauch. Sem. LOMI* **96** (1980), 237–239; English Transl.: *J. Soviet Math.* **21** (1983), 821–823.

[55] K. Pileckas, Recent advances in the theory of Stokes and Navier–Stokes equations in domains with non-compact boundaries, *Mathematical Theory in Fluid Mechanics*, 30–85, *Pitman Res. Notes Math. Ser.* **354**, Longman, Harlow, 1996.

[56] Y. Shibata, On the global existence of classical solutions of second order fully nonlinear hyperbolic equations with first order dissipation in the exterior domain, *Tsukuba J. Math.* **7** (1983), 1–68.

[57] Y. Shibata, On an exterior initial boundary value problem for Navier–Stokes equations, *Quart. Appl. Math.* **57** (1999), 117–155.

[58] Y. Shibata and S. Shimizu, A decay property of the Fourier transform and its application to the Stokes problem, *J. Math. Fluid Mech.* **3** (2001), 213–230.

[59] Y. Shimizu, $L^\infty$-estimate of first-order space derivatives of Stokes flow in a half space, *Funkcial. Ekvac.* **42** (1999), 291–309.

[60] C.G. Simader and H. Sohr, A new approach to the Helmholtz decomposition and the Neumann problem in $L^q$-spaces for bounded and exterior domains, *Mathematical Problems Relating to the Navier–Stokes Equation*, 1–35, *Ser. Adv. Math. Appl. Sci.* **11**, World Sci., NJ, 1992.

[61] V.A. Solonnikov, Estimates for solutions of nonstationary Navier–Stokes equations, *Zap. Nauch. Sem. LOMI* **38** (1973), 153–231; English Transl.: *J. Soviet Math.* **8** (1977), 467–528.

[62] V.A. Solonnikov, Stokes and Navier–Stokes equations in domains with noncompact boundaries, *Nonlinear PDE and Their Applications: College de France Seminar IV*, 240–349, *Res. Notes in Math.* **84**, Pitman, Boston, MA, 1983.

[63] V.A. Solonnikov, Boundary and initial-boundary value problems for the Navier–Stokes equations in domains with noncompact boundaries, *Mathematical Topics in Fluid Mechanics*, 117–162, *Pitman Res. Notes Math. Ser.* **274**, Longman, Harlow, 1992.

[64] H. Tanabe, *Equations of Evolution*, Pitman, London, 1979.

[65] S. Ukai, A solution formula for the Stokes equation in $\boldsymbol{R}^n_+$, *Commun. Pure Appl. Math.* **40** (1987), 611–621.

[66] M. Wiegner, Decay estimates for strong solutions of the Navier–Stokes equations in exterior domains, *Ann. Univ. Ferrara Sez. VII. Sc. Mat.* **46** (2000), 61–79.

Toshiaki Hishida
Fachbereich Mathematik
Technische Universität Darmstadt
D-64289 Darmstadt
Germany
e-mail: hishida@mathematik.tu-darmstadt.de

Advances in Mathematical Fluid Mechanics, 125–151


# Asymptotic Behavior at Infinity of Exterior Three-dimensional Steady Compressible Flow

T. Leonavičienė and K. Pileckas

**Abstract.** Steady compressible Navier–Stokes equations with zero velocity conditions at infinity are studied in a three-dimensional exterior domain. The case of small perturbations of large potential forces is considered. In order to solve the problem, a decomposition scheme is applied and the nonlinear problem is decomposed into three linear problems: Neumann-type problem, modified Stokes problem and transport equation. These linear problems are solved in weighted function spaces with detached asymptotics. The results on existence, uniqueness and asymptotics for the linearized problem and for the nonlinear problem are proved.

**Mathematics Subject Classification (2000).** 76N, 35Q.

**Keywords.** Steady compressible Navier–Stokes equations, three-dimensional exterior domain, weighted function spaces with detached asymptotics, pointwise decay at infinity.

## 1. Introduction

In this paper we study the asymptotic behavior of steady solutions to equations describing an isothermal motion of compressible viscous fluid (gas) in an exterior domain $\Omega \subset \mathbb{R}^3$. We assume that the flow domain $\Omega$ is an open, connected set, exterior to a compact set $B \subset \mathbb{R}^3$ with a sufficiently smooth boundary $\partial\Omega$. Moreover, we suppose that the interior of $B$ is non-empty and contains the origin of coordinates. In such a domain $\Omega$ we consider the classical Poisson–Stokes equations for unknown functions $\rho$ (density) and $\mathbf{v} = (v_1, v_2, v_3)$ (velocity):

$$\begin{cases} -\mu_1 \Delta \mathbf{v} - (\mu_1 + \mu_2)\nabla \mathrm{div}\, \mathbf{v} + \nabla \rho = \rho \mathbf{b} - \rho(\mathbf{v} \cdot \nabla)\mathbf{v}, & x \in \Omega, \\ \mathrm{div}\,(\rho \mathbf{v}) = 0, \quad x \in \Omega, \\ \mathbf{v} = 0, \quad x \in \partial\Omega. \end{cases} \tag{1.1}$$

This work was supported by Lithuanian State Science and Foundation grant A-524.

Here $\mu_1, \mu_2$ are constant coefficients of shear and bulk viscosities satisfying the conditions

$$\mu_1 > 0, \quad \mu_2 \geq -\frac{2}{3}\mu_1 \tag{1.2}$$

and $\mathbf{b}$ is a density of external forces. We assume that $\mathbf{b}$ is a "small" perturbation of a "large" potential force, i.e. that $\mathbf{b}$ has the form

$$\mathbf{b} = \nabla\Phi + \mathbf{f}, \tag{1.3}$$

where $\Phi$ is a potential which can be (as well as its derivatives) arbitrary "large" and $\mathbf{f}$ is a "small" perturbation of $\nabla\Phi$.

Since the flow domain $\Omega$ is unbounded, the equations (1.1) have to be supplied with the conditions at infinity. We assume that the velocity field $\mathbf{v}$ tends at infinity to zero and the density $\rho$ to a constant density and prescribe the following conditions:

$$\mathbf{v}(x) \to 0, \quad \rho(x) \to \rho_*, \quad \rho_* = \text{const} > 0, \quad |x| \to \infty. \tag{1.4}$$

It is a standard observation that the exact solution $(\rho_0, \mathbf{v}_0)$ of problem (1.1), (1.4) corresponding to the potential force $\nabla\Phi$ (i.e. $\mathbf{f} = 0$) is the rest state $(\rho_0, 0)$, where $\rho_0$ satisfies the equation

$$\nabla\rho_0 = \rho_0\nabla\Phi. \tag{1.5}$$

If $\Phi(x) \to 0$ as $|x| \to \infty$, from (1.4), (1.5) we find

$$\rho_0(x) = \rho_* \exp\Phi(x). \tag{1.6}$$

The equations for the perturbation $(\sigma = \rho - \rho_0, \mathbf{v})$ have the form

$$\begin{cases} -\mu_1\Delta\mathbf{v} - (\mu_1+\mu_2)\nabla\operatorname{div}\mathbf{v} + \nabla\sigma - \sigma\nabla\Phi = \mathbf{F}(\sigma,\mathbf{v}), & x \in \Omega, \\ \operatorname{div}(\rho_0\mathbf{v}) = -\operatorname{div}(\sigma\mathbf{v}), \quad x \in \Omega, \\ \mathbf{v} = 0, \quad x \in \partial\Omega, \\ \mathbf{v}(x) \to 0, \quad \sigma(x) \to 0, \quad |x| \to \infty, \end{cases} \tag{1.7}$$

where

$$F(\sigma,\mathbf{v}) = -(\rho_0+\sigma)(\mathbf{v}\cdot\nabla)\mathbf{v} + (\rho_0+\sigma)\mathbf{f}. \tag{1.8}$$

A possible linearization of the system (1.7) near the equilibrium state $(\rho_0, 0)$ reads

$$\begin{cases} -\mu_1\Delta\mathbf{v} - (\mu_1+\mu_2)\nabla\operatorname{div}\mathbf{v} + \nabla\sigma - \sigma\nabla\Phi = \mathbf{F}, \quad x \in \Omega, \\ \operatorname{div}(\rho_0\mathbf{v}) = -\operatorname{div}(\sigma\mathbf{w}), \quad x \in \Omega, \\ \mathbf{v} = 0, \quad x \in \partial\Omega, \quad \mathbf{v}(x) \to 0, \quad \sigma(x) \to 0, \quad |x| \to \infty, \end{cases} \tag{1.9}$$

where $\sigma$, $\mathbf{v}$ are unknown, while $\mathbf{F}$, $\mathbf{w}$ are given.

We will consider a slightly more general system

$$\begin{cases} -\mu_1\Delta\mathbf{v} - (\mu_1+\mu_2)\nabla\operatorname{div}\mathbf{v} + \nabla\sigma - \sigma\nabla\Phi = \mathbf{F}, \quad x \in \Omega, \\ \operatorname{div}(\rho_0\mathbf{v}) = -\operatorname{div}(\sigma\mathbf{w}) + g, \quad x \in \Omega, \\ \mathbf{v} = 0, \quad x \in \partial\Omega, \quad \mathbf{v}(x) \to 0, \quad \sigma(x) \to 0, \quad |x| \to \infty, \end{cases} \tag{1.9'}$$

where $g$ is given.

In order to solve the linearized problem (1.9′) and the nonlinear problem (1.7), we apply the decomposition scheme proposed in [15] (see also the references in [15] for the original application of the similar decomposition scheme to steady compressible Navier–Stokes equations linearized on a constant density). We represent the velocity field $\mathbf{v}$ as a sum

$$\mathbf{v} = \mathbf{u} + \nabla\varphi, \tag{1.10}$$

where

$$\operatorname{div}(\rho_0\mathbf{u}) = 0,\ x \in \Omega; \quad \mathbf{u}\cdot\mathbf{n} = 0,\ x \in \partial\Omega, \tag{1.11}$$

and

$$\frac{\partial\varphi}{\partial n} = 0, \quad x \in \partial\Omega. \tag{1.12}$$

As it is shown in [15], then the complicated mixed elliptic-hyperbolic system (1.9′) splits into the following three simpler problems:

$$\begin{cases} \Delta_{\rho_0}\varphi = -\operatorname{div}(\sigma\mathbf{w}) + g, & x \in \Omega, \\ \dfrac{\partial\varphi}{\partial n} = 0, \quad x \in \partial\Omega, \quad \varphi(x) \to 0, & |x| \to \infty, \end{cases} \tag{1.13}$$

$$\begin{cases} -\mu_1\Delta\mathbf{u} - (\mu_1+\mu_2)\nabla\operatorname{div}\mathbf{u} + \rho_0\nabla(\Pi/\rho_0) = \mathbf{G}, \quad x \in \Omega, \\ \operatorname{div}(\rho_0\mathbf{u}) = 0, \quad x \in \Omega, \\ \mathbf{u} = -\nabla\varphi, \quad x \in \partial\Omega, \\ \mathbf{u}(x) \to 0, \quad \Pi(x) \to 0, \quad |x| \to \infty, \end{cases} \tag{1.14}$$

$$\sigma + (2\mu_1+\mu_2)\operatorname{div}\Big(\frac{\sigma\mathbf{w}}{\rho_0}\Big) = \Pi, \quad x \in \Omega, \tag{1.15}$$

where $\Delta_{\rho_0} = \operatorname{div}\big(\rho_0(x)\nabla\big)$,

$$\begin{aligned} \mathbf{G} = \mathbf{F} &+ (2\mu_1+\mu_2)\nabla(\rho_0^{-2}\nabla\rho_0\cdot\mathbf{w}\sigma) - (2\mu_1+\mu_2)\nabla(\rho_0^{-1}\nabla\rho_0\cdot\nabla\varphi) \\ &- (2\mu_1+\mu_2)\rho_0^{-1}\nabla\rho_0\operatorname{div}\Big(\frac{\sigma\mathbf{w}}{\rho_0}\Big) + (2\mu_1+\mu_2)\nabla\Big(\frac{g}{\rho_0}\Big). \end{aligned} \tag{1.16}$$

Note that (1.13) and (1.14) are elliptic problem: (1.13) is a Neumann problem for the operator $\Delta_{\rho_0}$ and (1.14) is a Stokes type problem; the last problem (1.15) is the transport equation. Thus, the elliptic character of the original system (1.9′) is separated from its hyperbolic character. The hyperbolic effects are concentrated in the transport equation (1.15).

The solution $(\sigma, \mathbf{v})$ of the problem (1.9) can be found as follows. We define a linear map

$$\mathcal{L}\colon \tau \to \sigma$$

in the following way:

(a) for given $\tau$ we find $\varphi$ by solving the Neumann problem

$$\Delta_{\rho_0}\varphi = -\operatorname{div}(\tau\mathbf{w}),\ x \in \Omega; \quad \frac{\partial\varphi}{\partial n} = 0,\ x \in \partial\Omega;$$

(b) next we find the solution $(\mathbf{u}, \Pi)$ of the Stokes-type problem (1.14) taking in (1.16) $\sigma = \tau$;

(c) finally, we find $\sigma$ from the transport equation (1.15).

It is easy to verify that the fixed point $\sigma$ of the map $\mathcal{L}$ and the corresponding $(\mathbf{u}, \Pi, \varphi)$ solve the system (1.9′).

While the linearized system (1.9) is solved, we find the solution of the nonlinear problem (1.7) in the form $(\sigma,\ \mathbf{v} = \nabla\varphi + \mathbf{u})$, where $(\sigma, \varphi, \mathbf{u})$ is a fixed point of the nonlinear mapping

$$\mathcal{N}\colon (\tau, \xi, \mathbf{z}) \to (\sigma, \varphi, \mathbf{u}) \tag{1.17}$$

with $(\sigma, \varphi, \mathbf{u})$ being a solution of the linear problem (1.9) taking in it $\mathbf{F} = \mathbf{F}(\tau, \nabla\xi + \mathbf{z})$ and $\mathbf{w} = \nabla\xi + z$.

The solvability of problem (1.1)–(1.4) was studied in [14] using a different decomposition scheme and the techniques due to Matsumura, Nishida [2]. Moreover, in [14] were obtained decay estimates for the solution $(\rho = \rho_0 + \sigma,\ \mathbf{v})$ of (1.1)-(1.4). In particular, it is shown that the solution is "physically reasonable", i.e.

$$\mathbf{v}(x) = O\big(|x|^{-1}\big), \quad \nabla\mathbf{v}(x) = O\big(|x|^{-2}\big), \quad \sigma(x) = O\big(|x|^{-2}\big).$$

However, the analysis of [14], based on the integral representation formula for the Stokes problem and estimates of weakly singular integrals appearing in this representation, does not provide optimal decay rates for the higher order derivatives of the solution. Furthermore, with this method one cannot specify the asymptotic behavior of the solution as $|x| \to \infty$.

In this paper in order to investigate the solution of problem (1.1)–(1.4) we employ methods related with the application of weighted function spaces (e.g. [9, 5]). However, already the Navier–Stokes equations of the incompressible fluid motion are not solvable in classical weighted Sobolev (Hölder) spaces while considering the convective term $(\mathbf{v} \cdot \nabla)\mathbf{v}$ as a perturbation of the Stokes problem. On the other hand, if we study these equations in function spaces which elements take suitable asymptotic forms (i.e. the elements are defined as a sum of two parts one of which contains the main asymptotic term and another one belongs to the usual weighted space), then the Navier–Stokes problem is well-posed (see [6, 7]). We also refer to [3, 4, 11, 8] where such weighted spaces with detached asymptotics were successfully applied to study Navier–Stokes equations in other unbounded domains and to [10] were the spaces with detached asymptotic were applied to study viscoelastic flows in an exterior domain.

In this paper we prove that problem (1.1)–(1.4) is well-posed in appropriately constructed weighted Sobolev and Hölder spaces with detached asymptotics and obtain an accurate information about the asymptotic behavior of solutions.

The paper is organized as follows. In Section 2, we define the function spaces used in the paper and prove various embedding results for these spaces. In Section 3 we present known results concerning the Stokes problem in weighted spaces with detached asymptotics and study the Stokes-like problem (1.14) in such functional setting. In Section 4, we study other auxiliary problems ((1.15) and (1.13)) appearing in the decomposition scheme. Finally, in Section 5 and 6 we prove existence, uniqueness and assymptotics results for the linearized problem (1.9′) and the nonlinear problem (1.7) using the results of the previous sections and the fixed point argument. Let us point out that, although we consider here only the equations of an isothermal perfect gas, the result could be easily generalized to the case of barotropic motion.

The authors are deeply grateful to Prof. S. A. Nazarov for the helpful discussions.

## 2. Function spaces and auxiliary results

Here and in the sequel $\Omega \subset \mathbb{R}^3$ is an exterior domain with the smooth boundary $\partial\Omega$. We assume that the point $O \in \mathbb{R}^3 \backslash \Omega$. The outer normal to $\partial\Omega$ is denoted by $\mathbf{n}$. By $c$ we denote different constants. They may have different values in different formulae. If we want to fix the value of some constant, we regard it as $c_j$, $j = 1, 2, \ldots$. The norm of the element $u$ in the Banach space $V$ is denoted by $\|u; V\|$. We do not distinguish in notations between the spaces of scalar and vector valued functions. The difference is always clear from the context.

We use the following function spaces:

- $C^\infty(\Omega)$ is the set of all infinitely differentiable in $\bar{\Omega}$ functions; $C_0^\infty(\Omega)$ is the subset of functions from $C^\infty(\Omega)$ having compact supports in $\Omega$; $C_0^\infty(\bar{\Omega})$ is the set of functions from $C^\infty(\Omega)$ which are equal to zero in the neighborhood of infinity, i.e. for sufficiently large $|x|$ (but not necessary on $\partial\Omega$).
- $W^{l,q}(\Omega)$, $l \geq 0$, $q \in (1, \infty)$, is the usual Sobolev space and $\overset{\circ}{W}{}^{l,q}(\Omega)$ is the closure of $C_0^\infty(\Omega)$ in the norm $\|\cdot\,; W^{l,q}(\Omega)\|$; $L^q(\Omega) = W^{0,q}(\Omega)$.
- $W^{l-1/q,q}(\partial\Omega)$ is the space of traces on $\partial\Omega$ of functions from $W^{l,q}(\Omega)$, $l \geq 1$. The norm in $W^{l-1/q,q}(\partial\Omega)$ is defined by

$$\|\varphi; W^{l-1/q,q}(\partial\Omega)\| = \inf \|u; W^{l,q}(\Omega)\|$$

where the infimum is taken over all $u \in W^{l,q}(\Omega)$ such that $u = \varphi$ on $\partial\Omega$.
- Let $\mathbb{S}^2$ be a unit sphere in $\mathbb{R}^3$. $W^{l,q}(\mathbb{S}^2)$ (resp. $C^{l,\delta}(\mathbb{S}^2)$) is Sobolev (resp. Hölder) space of functions defined on $\mathbb{S}^2$.
- $D_0^1(\Omega)$ is the closure of $C_0^\infty(\Omega)$ in the Dirichlet norm $\|\nabla\cdot\,; L^2(\Omega)\|$. $D_0^{-1}(\Omega)$ is the dual space to $D_0^1(\Omega)$ with the usual duality norm.
- $V_\beta^{l,q}(\Omega)$, $l \geq 0$, $q \in (1, \infty)$, $\beta \in \mathbb{R}$, is the closure of $C_0^\infty(\bar{\Omega})$ in the weighted norm

$$\|u; V_\beta^{l,q}(\Omega)\| = \sum_{|\alpha| \leq l} \||x|^{\beta - l + |\alpha|} D^\alpha u;\, L^q(\Omega)\|,$$

where $\alpha = (\alpha_1, \alpha_2, \alpha_3)$, $D^\alpha = \partial^{|\alpha|}/\partial x_1^{\alpha_1} \partial x_2^{\alpha_2} \partial x_3^{\alpha_3}$, $\alpha_j \geq 0$, $|\alpha| = \alpha_1 + \alpha_2 + \alpha_3$.

- $\Lambda^{l,\delta}_{\beta}(\Omega)$ is the weighted Hölder space defined as the closure of $C_0^\infty(\bar{\Omega})$ in the norm

$$\|u;\Lambda^{l,\delta}_{\beta}(\Omega)\| = \sum_{|\alpha|\le l}\sup_{x\in\bar{\Omega}}\Big(|x|^{\beta-l-\delta+|\alpha|}\big|D^\alpha u(x)\big|\Big) + \sum_{|\alpha|=l}\sup_{x\in\Omega}\Big\{|x|^{\beta}\sup_{\substack{y\in\Omega\\ |x-y|<|x|/2}}\Big(|x-y|^{-\delta}\big|D^\alpha_x u(x)-D^\alpha_y u(y)\big|\Big)\Big\}.$$

- $\mathfrak{V}^{(l,s),q}_{\gamma,k}(\Omega)$, $l\ge 0$, $q\in(1,\infty)$, $s\ge l$, $k\in\mathbb{Z}$, $\gamma\in(l-3/q+k,\ l+1-3/q+k)$, is the weighted Sobolev space with detached asymptotics, i.e. the space of functions admiting the asymptotic representation

$$u(x) = r^{-k}\mathfrak{U}(\theta)+\widetilde{u}(x), \tag{2.1}$$

where $r=|x|$, $\theta\in\mathbb{S}^2$, $U\in W^{l,q}(\mathbb{S}^2)$, $\widetilde{u}\in V^{l,q}_{\gamma}(\Omega)$. The norm in $\mathfrak{V}^{(l,s),q}_{\gamma,k}(\Omega)$ is defined by

$$\|u;\mathfrak{V}^{(l,s),q}_{\gamma,k}(\Omega)\| = \|\mathfrak{U};W^{s,q}(\mathbb{S}^2)\|+\|\widetilde{u};V^{l,q}_{\gamma}(\Omega)\|.$$

*Functions* $\mathfrak{U}$ *and* $\widetilde{u}$ *are called the attributes of* $u\in\mathfrak{V}^{(l,s),q}_{\gamma,k}(\Omega)$.

- $\mathfrak{C}^{(l,s),\delta}_{\gamma,k}(\Omega)$ is the weighted Hölder space of functions with detached asymptotics (2.1) having the attributes $\mathfrak{U}\in C^{s,\delta}(\mathbb{S}^2)$, $\widetilde{u}\in\Lambda^{l,\delta}_{\gamma}(\Omega)$, where $\gamma\in(l+\delta+k,\ l+\delta+1+k)$, $\delta\in(0,1)$. The norm in $\mathfrak{C}^{(l,s)\delta}_{\gamma,k}(\Omega)$ is given by the formula

$$\|u;\mathfrak{C}^{(l,s)\delta}_{\gamma,k}(\Omega)\| = \|\mathfrak{U};C^{s,\delta}(\mathbb{S}^2)\|+\|\widetilde{u};\Lambda^{l,\delta}_{\beta}(\Omega)\|.$$

From the definitions of the norms follows

**Lemma 2.1.** (i) *Let* $u\in V^{l,q}_{\beta}(\Omega)$ $\big(u\in\Lambda^{l,\delta}_{\beta}(\Omega)\big)$, $m\le l$, $|\alpha|\le l$. *Then* $u\in V^{m,q}_{\beta-l+m}(\Omega)$, $D^\alpha u\in V^{l-|\alpha|,q}_{\beta}(\Omega)$ $\big(u\in\Lambda^{m,\delta}_{\beta-l+m}(\Omega),\ D^\alpha u\in\Lambda^{l-|\alpha|,\delta}_{\beta}(\Omega)\big)$.

(ii) *Let* $u\in\mathfrak{V}^{(l,s),q}_{\gamma,k}(\Omega)$ $\big(u\in\mathfrak{C}^{(l,s),\delta}_{\gamma,k}(\Omega)\big)$, $m\le l$, $m\le p\le s$, $|\alpha|\le l$. *Then* $u\in\mathfrak{V}^{(m,p),q}_{\gamma-l+m,k}(\Omega)$, $D^\alpha u\in\mathfrak{V}^{(l-|\alpha|,s-|\alpha|),q}_{\gamma,k+|\alpha|}(\Omega)$ $\big(u\in\mathfrak{C}^{(m,p),\delta}_{\gamma-l+m,k}(\Omega),\ D^\alpha u\in\mathfrak{C}^{(l-|\alpha|,s-|\alpha|),\delta}_{\gamma,k+|\alpha|}(\Omega)\big)$.

The following embedding results are proved in [1].

**Lemma 2.2.** *Let* $u\in V^{l,q}_{\beta}(\Omega)$.

(i) *If* $ql\le 3$ *with* $q\le s\le 3q/(3-ql)$, *then* $u\in V^{0,s}_{\beta-l-3/s+3/q}(\Omega)$ *and*

$$\|u;V^{0,s}_{\beta-l-3/s+3/q}(\Omega)\|\le c\,\|u;V^{l,q}_{\beta}(\Omega)\|. \tag{2.2}$$

(ii) *If* $ql>3$ *and* $m+\delta\le l-3/q$ *with* $\delta\in(0,1)$, *then* $u\in\Lambda^{m,\delta}_{m+\delta+\beta-l+3/q}(\Omega)$ *and*

$$\|u;\Lambda^{m,\delta}_{m+\delta+\beta-l+3/q}(\Omega)\|\le c\,\|u;V^{l,q}_{\beta}(\Omega)\|. \tag{2.3}$$

**Lemma 2.3.** *Let* $v \in \Lambda^{l+1,\delta}_{l+1+\delta+\beta_1}(\Omega)$.

(i) *If* $u \in V^{m,q}_{\beta_2}(\Omega)$, $m \le l+1$, *then* $vu \in V^{m,q}_{\beta_1+\beta_2}(\Omega)$ *and*

$$\|vu; V^{m,q}_{\beta_1+\beta_2}(\Omega)\| \le c\,\|v; \Lambda^{l+1,\delta}_{l+1+\delta+\beta_1}(\Omega)\|\|u; V^{m,q}_{\beta_2}(\Omega)\|. \tag{2.4}$$

(ii) *If* $u \in \Lambda^{m,\delta}_{\beta_2}(\Omega)$, $m \le l+1$, *then* $vu \in \Lambda^{m,\delta}_{\beta_1+\beta_2}(\Omega)$ *and*

$$\|vu; \Lambda^{m,\delta}_{\beta_1+\beta_2}(\Omega)\| \le c\,\|v; \Lambda^{l+1,\delta}_{l+1+\delta+\beta_1}(\Omega)\|\|u; \Lambda^{m,\delta}_{\beta_2}(\Omega)\|. \tag{2.5}$$

*Proof.* Since

$$D^\alpha(vu) = \sum_{\substack{\mu_j \le \alpha_j \\ j=1,2,3}} \binom{\alpha}{\mu} D^{\alpha-\mu} v D^\mu u, \quad \binom{\alpha}{\mu} = \frac{\alpha}{\mu(\alpha-\mu)}, \qquad \alpha! = \alpha_1!\alpha_2!\alpha_3!,$$

we get

$$\begin{aligned}
&\int_\Omega |D^\alpha(vu)|^q |x|^{q(\beta_2+\beta_1-m+|\alpha|)}\,dx \\
&\le c \sum_{|\mu|=0}^{|\alpha|} \int_\Omega |x|^{q(\beta_1+|\alpha|-|\mu|)} |D^{\alpha-\mu} v|^q |x|^{q(\beta_2-m+|\mu|)} |D^\mu u|^q\,dx \\
&\le c\,\|v; \Lambda^{l+1,\delta}_{\beta_1+l+1+\delta}(\Omega)\|\|u; V^{m,q}_{\beta_2}(\Omega)\|.
\end{aligned}$$

Summing these inequalities over $\alpha$ with $|\alpha| \le m$, we derive (2.4). The estimate (2.5) can be proved analogously. □

Let us formulate other results which give the possibility to estimate nonlinear terms in (1.8), (1.16).

**Lemma 2.4.** (i) *Let* $\mathbf{v} \in \mathfrak{V}^{(l+1,s+1),q}_{\gamma,1}(\Omega)$ *with* $l \ge 1$, $s \ge l$, $3/2 < q < \infty$ *and* $\gamma \in (l+2-3/q,\, l+3-3/q)$. *Then* $\mathbf{v} \otimes \mathbf{v} \in \mathfrak{V}^{(l,s),q}_{\gamma,2}(\Omega)$ *and*

$$\|\mathbf{v} \otimes \mathbf{v}; \mathfrak{V}^{(l,s),q}_{\gamma,2}(\Omega)\| \le c\,\|\mathbf{v}; \mathfrak{V}^{(l+1,s+1),q}_{\gamma,1}(\Omega)\|^2. \tag{2.6}$$

(ii) *Let* $\mathbf{v} \in \mathfrak{C}^{(l+1,s+1),\delta}_{\gamma,1}(\Omega)$, $l \ge 1$, $s \ge l$, $\delta \in (0,1)$, $\gamma \in (l+2+\delta,\, l+3+\delta)$. *Then* $\mathbf{v} \otimes \mathbf{v} \in \mathfrak{C}^{(l,s),\delta}_{\gamma,2}(\Omega)$ *and*

$$\|\mathbf{v} \otimes \mathbf{v}; \mathfrak{C}^{(l,s),\delta}_{\gamma,2}(\Omega)\| \le c\,\|\mathbf{v}; \mathfrak{C}^{(l+1,s+1),\delta}_{\gamma,1}(\Omega)\|^2. \tag{2.7}$$

**Lemma 2.5.** (i) *Let* $g \in \mathfrak{V}^{(l+1,s),q}_{\gamma,k-1}(\Omega)$, $h \in \mathfrak{V}^{(l,s),q}_{\gamma,k}(\Omega)$, $l \ge 2$, $s \ge l+1$, $q > 3/2$, $\gamma \in (l+k-3/q,\, l+k+1-3/q)$, $k \ge 2$. *Then* $gh \in V^{l,q}_\gamma(\Omega)$ *and there holds the estimate*

$$\|gh; V^{l,q}_\gamma(\Omega)\| \le c\,\|g; \mathfrak{V}^{(l+1,s),q}_{\gamma,k-1}(\Omega)\|\|h; \mathfrak{V}^{(l,s),q}_{\gamma,k}(\Omega)\|. \tag{2.8}$$

(ii) *Let* $g \in \mathfrak{C}^{(l+1,s),\delta}_{\gamma,k-1}(\Omega)$, $h \in \mathfrak{C}^{(l,s),\delta}_{\gamma,k}(\Omega)$, $l \ge 0$, $s \ge l+1$, $\delta \in (0,1)$, $\gamma \in (l+k+\delta,\, l+k+1+\delta)$, $k \ge 2$. *Then* $gh \in \Lambda^{l,\delta}_\gamma(\Omega)$ *and*

$$\|gh; \Lambda^{l,\delta}_\gamma(\Omega)\| \le c\,\|g; \mathfrak{C}^{(l+1,s),\delta}_{\gamma,k-1}(\Omega)\|\|h; \mathfrak{C}^{(l,s),\delta}_{\gamma,k}(\Omega)\|. \tag{2.9}$$

Estimate (2.6) is proved in [10] for $s = l$ (see Lemma 3.3 in [10]). Estimate (2.8) is proved in [10] for $s = l+1$, $k = 2$ (see Lemma 3.5). The proof for arbitrary $s \geq l+1$ and $k \geq 2$ is completely analogous. Estimates (2.7), (2.9) easily follow from the definition of the norm in the space $\mathfrak{C}^{(l,s),\delta}_{\gamma,k}(\Omega)$.

## 3. Stokes and modified Stokes problems in weighted spaces

### 3.1. Stokes problem in weighted $L^q$ and Hölder spaces

Let us consider in $\Omega$ the Stokes problem

$$\begin{cases} -\nu\Delta \mathbf{u} + \nabla p = \mathbf{f} & \text{in } \Omega, \\ \operatorname{div} \mathbf{u} = g & \text{in } \Omega, \\ \mathbf{u} = \mathbf{h} & \text{on } \partial\Omega. \end{cases} \tag{3.1}$$

We associate with problem (3.1) the mapping $S^{l,q}_{\beta} \colon \mathcal{D}^{l,q}_{\beta}V(\Omega) \to \mathcal{R}^{l,q}_{\beta}V(\Omega)$ defined by

$$(\mathbf{u}, p) \to (\mathbf{f}, g, \mathbf{h}) = S^{l,q}_{\beta}(\mathbf{u}, p), \tag{3.2}$$

where

$$\begin{aligned} &\mathcal{D}^{l,q}_{\beta}V(\Omega) \equiv V^{l+1,q}_{\beta}(\Omega) \times V^{l,q}_{\beta}(\Omega), \\ &\mathcal{R}^{l,q}_{\beta}V(\Omega) \equiv V^{l-1,q}_{\beta}(\Omega) \times V^{l,q}_{\beta}(\Omega) \times W^{l+1-1/q,q}(\partial\Omega), \\ &l \geq 1, \quad q \in (1,\infty), \quad \beta \in \mathbb{R}. \end{aligned}$$

The following results are well known (see [9, 6]).

**Theorem 3.1.** (i) *If*

$$\beta \in (l+1-3/q,\, l+2-3/q), \tag{3.3}$$

*then the mapping* (3.2) *is an isomorphism.*

(ii) *Let* $(\mathbf{f}, g, \mathbf{h}) \in \mathcal{R}^{l,q}_{\gamma}V(\Omega) \subset \mathcal{R}^{l,q}_{\beta}V(\Omega)$ *with*

$$\gamma \in (l+2-3/q,\, l+3-3/q). \tag{3.4}$$

*Then the solution* $(\mathbf{u}, p) \in \mathcal{D}^{l,q}_{\beta}V(\Omega)$ *admits the asymptotic representation*

$$(\mathbf{u}, p) = (\mathbf{u}^0, p^0) + (\widetilde{\mathbf{u}}, \widetilde{p}), \tag{3.5}$$

*where* $(\widetilde{\mathbf{u}}, \widetilde{p}) \in \mathcal{D}^{l,q}_{\gamma}V(\Omega)$ *and*

$$(\mathbf{u}^0, p^0) = b_1\mathbf{E}^{(1)} + b_2\mathbf{E}^{(2)} + b_3\mathbf{E}^{(3)}, \tag{3.6}$$

*with* $\mathbf{E}^{(j)}$ *denoting the* $j$*-th column of the fundamental matrix for the Stokes operator in* $\mathbb{R}^3$ *and* $b_j \in \mathbb{R}$, $j = 1,2,3$. *Moreover, there holds the estimate*

$$\big\|(\widetilde{\mathbf{u}}, \widetilde{p}); \mathcal{D}^{l,q}_{\gamma}V(\Omega)\big\| + |b_1| + |b_2| + |b_3| \leq c\,\big\|(\mathbf{f}, g, \mathbf{h}); \mathcal{R}^{l,q}_{\gamma}V(\Omega)\big\|. \tag{3.7}$$

*Remark* 3.2. The columns of the fundamental matrix for the Stokes operator in $\mathbb{R}^3$ are defined by

$$\mathbf{E}^{(j)}(x)=\frac{1}{8\pi\nu|x|^3}\left(\delta_{j1}|x|^2+x_1x_j,\ \delta_{j2}|x|^2+x_2x_j,\ \delta_{j3}|x|^2+x_3x_j,\ 2\nu x_j\right)^{\mathsf{T}},\ j=1,2,3.$$

Let us consider the problem (3.1) in weighted Sobolev spaces with detached asymptotics. Denote

$$\mathcal{D}^{l,q}_{\gamma}\mathfrak{V}(\Omega) \equiv \mathfrak{V}^{(l+1,l+2),q}_{\gamma,1}(\Omega) \times \mathfrak{V}^{l,l+1}_{\gamma,2}(\Omega),$$
$$\mathcal{R}^{l,q}_{\gamma}\mathfrak{V}(\Omega) \equiv \mathfrak{V}^{(l-1,l),q}_{\gamma,3}(\Omega) \times \mathfrak{V}^{(l,l+1),q}_{\gamma,2}(\Omega) \times W^{l+1-1/q,q}(\partial\Omega)$$

with $\gamma$ satisfying (3.4). It is not difficult to compute that

$$\mathcal{D}^{l,q}_{\gamma}\mathfrak{V}(\Omega) \subset \mathcal{D}^{l,q}_{\beta}V(\Omega), \quad \mathcal{R}^{l,q}_{\gamma}\mathfrak{V}(\Omega) \subset \mathcal{R}^{l,q}_{\beta}V(\Omega)$$

with $\beta$ taken from (3.3).

Let $\mathfrak{S}^{l,q}_{\gamma}$ be the operator of the Stokes problem (3.1) acting on the domain $\mathcal{D}^{l,q}_{\gamma}\mathfrak{V}(\Omega)$ to the range $\mathcal{R}^{l,q}_{\gamma}\mathfrak{V}(\Omega)$.

**Theorem 3.3.** *(see [6, 7].) Let $(\mathbf{f}, g, \mathbf{h}) \in \mathcal{R}^{l,q}_{\gamma}\mathfrak{V}(\Omega)$. Problem (3.1) has a solution $(\mathbf{u}, p) \in \mathcal{D}^{l,q}_{\gamma}\mathfrak{V}(\Omega)$ if and only if there holds the compatibility condition*

$$\int\limits_{\mathbb{S}^2} \mathfrak{F}(\theta)ds_\theta = 0, \tag{3.8}$$

*where $\mathfrak{F}(\theta)$, $\widetilde{\mathbf{f}}$ are the attributes of the function $\mathbf{f}$ in representation (2.1). The solution is unique and there holds the estimate*

$$\big\|(\mathbf{u},p);\ \mathcal{D}^{l,q}_{\gamma}\mathfrak{V}(\Omega)\big\| \le c\,\big\|(\mathbf{f},g,\mathbf{h});\ \mathcal{R}^{l,q}_{\gamma}\mathfrak{V}(\Omega)\big\|. \tag{3.9}$$

The analogous results are also valid in weighted Hölder spaces. Let us fix $l \ge 1$, $\delta \in (0,1)$, $\beta \in (l+1+\delta,\, l+2+\delta)$, $\gamma \in (l+2+\delta,\, l+3+\delta)$ and define the spaces

$$\mathcal{D}^{l,\delta}_{\beta}\Lambda(\Omega) \equiv \Lambda^{l+1,\delta}_{\beta}(\Omega) \times \Lambda^{l,\delta}_{\beta}(\Omega),$$
$$\mathcal{R}^{l,\delta}_{\beta}\Lambda(\Omega) \equiv \Lambda^{l-1,\delta}_{\beta}(\Omega) \times \Lambda^{l,\delta}_{\beta}(\Omega) \times C^{l+1,\delta}(\partial\Omega),$$
$$\mathcal{D}^{l,\delta}_{\gamma}\mathfrak{C}(\Omega) \equiv \mathfrak{C}^{(l+1,l+2)\delta}_{\gamma,1}(\Omega) \times \mathfrak{C}^{(l,l+1)\delta}_{\gamma,2}(\Omega),$$
$$\mathcal{R}^{l,\delta}_{\gamma}\mathfrak{C}(\Omega) \equiv \mathfrak{C}^{(l-1,l)\delta}_{\gamma,3}(\Omega) \times \mathfrak{C}^{(l,l+1)\delta}_{\gamma,2}(\Omega) \times C^{l+1,\delta}(\partial\Omega).$$

**Theorem 3.4.** *(see [6].) Let $(\mathbf{f}, g, \mathbf{h}) \in \mathcal{R}^{l,\delta}_{\gamma}\mathfrak{C}(\Omega)$. Problem (3.1) has a solution $(\mathbf{u}, p) \in \mathcal{D}^{l,\delta}_{\gamma}\mathfrak{C}(\Omega)$ if and only if there holds compatibility condition (3.8). The solution is unique and*

$$\big\|(\mathbf{u},p);\ \mathcal{D}^{l,\delta}_{\gamma}\mathfrak{C}(\Omega)\big\| \le c\,\big\|(\mathbf{f},g,\mathbf{h});\ \mathcal{R}^{l,\delta}_{\gamma}\mathfrak{C}(\Omega)\big\|. \tag{3.10}$$

**3.2. Modified Stokes problem**

Let us consider the problem

$$\begin{cases} -\mu_1 \Delta \mathbf{u} - (\mu_1 + \mu_2)\nabla \mathrm{div}\, \mathbf{u} + \rho_0 \nabla(\Pi/\rho_0) = \mathbf{f} \quad \text{in } \Omega, \\ \mathrm{div}\,(\rho_0 \mathbf{u}) = g \quad \text{in } \Omega, \\ \mathbf{u} = \mathbf{h} \quad \text{on } \partial\Omega, \end{cases} \tag{3.11}$$

where $\rho_0(x) = \rho_* \exp \Phi(x)$, $\Phi(x) \in \Lambda^{l+1,\delta}_{l+1+\delta+\gamma_0}(\Omega)$ with $\gamma_0 > 1$.

First we deal with the case $\mathbf{h} = 0$ which we regard as the "homogeneous" problem (3.11) and denote $(3.11)_0$.

By weak solution of problem $(3.11)_0$ we understand a pair $(\mathbf{u}, \Pi) \in D_0^1(\Omega) \times L^2(\Omega)$ satisfying the integral identity

$$\begin{aligned} &\mu_1 \int_\Omega \nabla \mathbf{u} : \nabla \boldsymbol{\eta}\, dx + (\mu_1 + \mu_2) \int_\Omega \mathrm{div}\, \mathbf{u}\, \mathrm{div}\, \boldsymbol{\eta}\, dx \\ &- \int_\Omega \rho_0^{-1} \Pi\, \mathrm{div}\,(\rho_0 \boldsymbol{\eta})\, dx = \int_\Omega \mathbf{f} \cdot \boldsymbol{\eta}\, dx, \quad \forall \boldsymbol{\eta} \in D_0^1(\Omega). \end{aligned} \tag{3.12}$$

and the equation $\mathrm{div}\,(\rho_0 \mathbf{u}) = g$.

**Theorem 3.5.** *Let $\Omega$ be an exterior domain with Lipschitz boundary. Assume that $\mathbf{f} \in D_0^{-1}(\Omega)$, $g \in L^2(\Omega)$. Then there exists a unique weak solution $(\mathbf{u}, \Pi)$ of problem $(3.11)_0$ and there holds the estimate*

$$\|\mathbf{u}; D_0^1(\Omega)\| + \|\Pi; L^2(\Omega)\| \le c\Big(\|\mathbf{f}; D_0^{-1}(\Omega)\| + \|g; L^2(\Omega)\|\Big). \tag{3.13}$$

The proof of Theorem 3.5 is standard and does not differ from the proof of the analogous result in the case of a bounded domain $\Omega$ (see [15]). Note that the proof is based on the following.

**Lemma 3.6.** (i) *Let $\mathbf{u} \in D_0^1(\Omega)$, $g = \mathrm{div}\,(\rho_0 \mathbf{u})$. Then $g \in L^2(\Omega)$ and*

$$\|g; L^2(\Omega)\| \le c\,\|\mathbf{u}; D_0^1(\Omega)\|. \tag{3.14}$$

(ii) *Every function $g \in L^2(\Omega)$ can be represented in the form $g = \mathrm{div}\,(\rho_0 \mathbf{u})$, $\mathbf{u} \in D_0^1(\Omega)$ and there holds the estimate*

$$\|\mathbf{u}; D_0^1(\Omega)\| \le c\,\|g; L^2(\Omega)\|. \tag{3.15}$$

*Proof.* (i) Since $\rho_0^{-1} \nabla \rho_0 = \nabla \Phi$, we have $g = \mathrm{div}\,(\rho_0 \mathbf{u}) = \rho_* \exp \Phi\, (\nabla \Phi \cdot \mathbf{u} + \mathrm{div}\, \mathbf{u})$ and using the inclusion $\nabla \Phi \in \Lambda^{l+1,\delta}_{l+1+\delta+\gamma_0}(\Omega)$, we derive the estimate (3.14).

(ii) Let $\mathbf{v} \in D_0^1(\Omega)$ be a solution of the problem

$$\begin{cases} \mathrm{div}\, \mathbf{v} = g, & x \in \Omega, \\ \mathbf{v} = 0, & x \in \partial\Omega, \end{cases}$$

satisfying the estimate

$$\|\mathbf{v}; D_0^1(\Omega)\| \le c\,\|g; L^2(\Omega)\|.$$

Such a solution $\mathbf{v}$ exists for each $g \in L^2(\Omega)$ (see [16]). Taking $\mathbf{u} = \rho_0^{-1}\mathbf{v}$ we get $\operatorname{div}(\rho_0\mathbf{u}) = g$ and the estimate (3.15) holds true. □

Let us consider problem $(3.11)_0$ in weighted function spaces.

**Theorem 3.7.** (i) *Let* $\mathbf{f} \in V_\beta^{l-1,q}(\Omega)$, $g \in V_\beta^{l,q}(\Omega)$, $l \geq 1$, $q \geq 6/5$, $\beta \in (l + 3/2 - 3/q, l + 2 - 3/q)$. *Then problem* $(3.11)_0$ *has a unique solution* $(\mathbf{u}, \Pi) \in V_\beta^{l+1,q}(\Omega) \times V_\beta^{l,q}(\Omega)$ *satisfying the estimate*

$$\|(\mathbf{u}, \Pi); V_\beta^{l+1,q}(\Omega) \times V_\beta^{l,q}(\Omega)\| \leq c\|(\mathbf{f}, g); V_\beta^{l-1,q}(\Omega) \times V_\beta^{l,q}(\Omega)\|. \tag{3.16}$$

(ii) *Let* $(\mathbf{f}, g) \in \Lambda_\beta^{l-1,\delta}(\Omega) \times \Lambda_\beta^{l,\delta}(\Omega)$, $l \geq 1$, $\delta \in (0,1)$, $\beta \in (l+\delta+3/2, l+\delta+2)$. *Then problem* $(3.11)_0$ *has a unique solution* $(\mathbf{u}, \Pi) \in \Lambda_\beta^{l+1,\delta}(\Omega) \times \Lambda_\beta^{l,\delta}(\Omega)$ *and*

$$\|(\mathbf{u}, \Pi); \Lambda_\beta^{l+1,\delta}(\Omega) \times \Lambda_\beta^{l,\delta}(\Omega)\| \leq c\|(\mathbf{f}, g); \Lambda_\beta^{l-1,\delta}(\Omega) \times \Lambda_\beta^{l,\delta}(\Omega)\|. \tag{3.17}$$

*Proof.* (i) Let first consider the case $q = 2$. For each $\mathbf{u} \in D_0^1(\Omega)$ there holds the estimate

$$\int_\Omega |x|^{-2}|\mathbf{u}|^2\, dx \leq c\int_\Omega |\nabla\mathbf{u}|^2\, dx.$$

Hence, $D_0^1(\Omega) = V_0^{1,2}(\Omega)$. Moreover, if $\beta \geq l$, then

$$\Big|\int_\Omega \mathbf{f}\cdot\boldsymbol{\eta}\, dx\Big| \leq \Big(\int_\Omega |x|^2|\mathbf{f}|^2\, dx\Big)^{1/2}\Big(\int_\Omega |x|^{-2}|\boldsymbol{\eta}|^2\, dx\Big)^{1/2}$$
$$\leq c\Big(\int_\Omega |x|^{2(\beta+1-l)}|\mathbf{f}|^2\, dx\Big)^{1/2}\Big(\int_\Omega |\nabla\boldsymbol{\eta}|^2\, dx\Big)^{1/2}, \quad \forall\boldsymbol{\eta} \in D_0^1(\Omega).$$

Therefore, for $\beta \geq l$, $V_\beta^{l-1,2}(\Omega) \subset D_0^{-1}(\Omega)$ and problem $(3.11)_0$ has a unique weak solution $(\mathbf{u}, \Pi) \in D_0^1(\Omega) \times L^2(\Omega) = V_0^{1,2}(\Omega) \times V_0^{0,2}(\Omega)$. From results of the paper [15] it follows that $(\mathbf{u}, \Pi) \in W_{\mathrm{loc}}^{2,2}(\Omega) \times W_{\mathrm{loc}}^{1,2}(\Omega)$ and equations $(3.11)_0$ are satisfied almost everywhere in $\Omega$. We rewrite $(3.11)_0$ in the form

$$\begin{cases} -\mu_1\Delta\mathbf{u} + \nabla\Pi = \mathbf{f} + \rho_0^{-1}\nabla\rho_0\Pi - (\mu_1+\mu_2)\nabla(\rho_0^{-1}\nabla\rho_0\cdot\mathbf{u}), & x \in \Omega, \\ \operatorname{div}\mathbf{u} = g - \rho_0^{-1}\nabla\rho_0\cdot\mathbf{u}, \quad x \in \Omega, \\ \mathbf{u} = 0, \quad x \in \partial\Omega, \end{cases} \tag{3.18}$$

and consider $(\mathbf{u}, \Pi)$ as a solution of Stokes problem (3.1) with the right-hand side $(\mathbf{F}, G, 0)$, where

$$\begin{aligned} \mathbf{F} &= \mathbf{f} + \Pi\nabla\Phi - (\mu_1+\mu_2)\nabla(\nabla\Phi\cdot\mathbf{u}), \\ G &= g - \nabla\Phi\cdot\mathbf{u} \end{aligned} \tag{3.19}$$

(remember that $\nabla\Phi = \rho_0^{-1}\nabla\rho_0$). By Lemma 2.3 we get $\nabla\Phi\cdot\mathbf{u} \in V_{\gamma_0}^{1,2}(\Omega)$, $\Pi\nabla\Phi \in V_{\gamma_0}^{0,2}(\Omega)$. Since $\mathbf{f} \in V_\beta^{l-1,2}(\Omega) \subset V_{\beta-l+1}^{0,2}(\Omega)$, $g \in V_\beta^{l,2}(\Omega) \subset V_{\beta-l+1}^{1,2}(\Omega)$, we conclude

$$\mathbf{F} \in V_{\beta_1}^{0,2}(\Omega), \quad G \in V_{\beta_1}^{1,2}(\Omega),$$

with $\beta_1 = \min(\gamma_0, \beta-l+1)$. For $\gamma_0 > 1$, $\beta \in (l, l+1/2)$ we have $\beta_1 - 1 \in (0,1/2)$ and in virtue of Theorem 3.1 *(i)* the solution $(\mathbf{u}, \Pi)$ of Stokes problem (3.18) belongs to the space $\mathcal{D}^{1,2}_{\beta_1}V(\Omega) = V^{2,2}_{\beta_1}(\Omega) \times V^{1,2}_{\beta_1}(\Omega)$. Applying again Lemma 2.3, we get $\nabla\Phi \cdot \mathbf{u} \in V^{2,2}_{\beta_1+\gamma_0}(\Omega)$, $\Pi\nabla\Phi \in V^{1,2}_{\beta_1+\gamma_0}(\Omega)$, so that

$$\mathbf{F} \in V^{1,2}_{\beta_2}(\Omega), \quad G \in V^{2,2}_{\beta_2}(\Omega)$$

with $\beta_2 = \min(2\gamma_0,\, \beta - l + 2)$. We have $\beta_2 - 2 \in (0,1/2)$ and Theorem 3.1 (i) implies $(\mathbf{u}, \Pi) \in \mathcal{D}^{2,2}_{\beta_2}V(\Omega)$. Repeating this argument $l$ times we derive

$$(\mathbf{F}, G) \in V^{l-1,2}_{\beta_l}(\Omega) \times V^{l,2}_{\beta_l}(\Omega)$$

with $\beta_l = \min(l\gamma_0, \beta) \in (l, l+1/2)$ and conclude that $(\mathbf{u}, \Pi) \in \mathcal{D}^{l,2}_{\beta_l}V(\Omega)$. If $l\gamma_0 \geq \beta$, then $\mathcal{D}^{l,2}_{\beta_l}V(\Omega) \subset \mathcal{D}^{l,2}_{\beta}V(\Omega)$ and in the case $q = 2$ the theorem is proved. If $l\gamma_0 < \beta$, we continue the iteration process:

$$\begin{aligned}
&(\mathbf{u}, \Pi) \in \mathcal{D}^{l,2}_{\beta_l}V(\Omega) \\
&\overset{\text{Lemma 2.3}}{\Longrightarrow} (\mathbf{F}, G) \in V^{l-1,2}_{\beta_{l+1}}(\Omega) \times V^{l,2}_{\beta_{l+1}}(\Omega), \; \beta_{l+1} = \min\big((l+1)\gamma_0, \beta\big) \\
&\overset{\text{Theorem 3.1 (i)}}{\Longrightarrow} (\mathbf{u}, \Pi) \in \mathcal{D}^{l,2}_{\beta_{l+1}}V(\Omega) \\
&\Longrightarrow \ldots \overset{\text{Lemma 2.3}}{\Longrightarrow} (\mathbf{F}, G) \in V^{l-1,2}_{\beta_{l+m}}(\Omega) \times V^{l,2}_{\beta_{l+m}}(\Omega) \\
&\overset{\text{Theorem 3.1 (i)}}{\Longrightarrow} (\mathbf{u}, \Pi) \in \mathcal{D}^{l,2}_{\beta_{l+m}}V(\Omega), \quad \beta_{l+m} = \min\big((l+m)\gamma_0, \beta\big).
\end{aligned}$$

Taking $m$ such that $(l+m)\gamma_0 \geq \beta$, we obtain $(\mathbf{u}, \Pi) \in \mathcal{D}^{l,2}_{\beta}V(\Omega)$. Supplying mentioned above inclusions by the corresponding estimates we get for $(\mathbf{u}, \Pi)$ the inequality (3.16) at $q = 2$.

Let us consider the case $q \in [6/5, \infty)$. If $q \in [6/5, 2]$, $\beta \geq l + 3/2 - 3/q$, we have

$$\Big|\int\limits_\Omega \mathbf{f} \cdot \boldsymbol{\eta}\, \mathrm{d}x\Big| \leq \Big(\int\limits_\Omega |\mathbf{f}|^q |x|^{q(\beta-l+1)}\, \mathrm{d}x\Big)^{1/q} \Big(\int\limits_\Omega |\boldsymbol{\eta}|^{q'} |x|^{q'(-\beta+l-1)}\, \mathrm{d}x\Big)^{1/q'},$$

where $1/q + 1/q' = 1$. Since $q' \in [2,6]$, in view of Lemma 2.2 (i) there hold the embeddings $D^1_0(\Omega) = V^{1,2}_0(\Omega) \subset V^{0,q'}_{1/2-3/q'}(\Omega) \subset V^{0,q'}_{-\beta+l-1}(\Omega)$. Therefore,

$$\Big|\int\limits_\Omega \mathbf{f} \cdot \boldsymbol{\eta}\, \mathrm{d}x\Big| \leq c\,\big\|\mathbf{f}; V^{0,q}_{\beta-l+1}(\Omega)\big\|\big\|\boldsymbol{\eta}; D^1_0(\Omega)\big\|.$$

If $q > 2$, $\beta > l + 3/2 - 3/q$, then

$$\begin{aligned}
\Big|\int\limits_\Omega \mathbf{f} \cdot \boldsymbol{\eta}\, \mathrm{d}x\Big| &\leq \Big(\int\limits_\Omega |\mathbf{f}|^q |x|^{q(\beta-l+1)}\, \mathrm{d}x\Big)^{1/q} \Big(\int\limits_\Omega |\boldsymbol{\eta}|^6\, \mathrm{d}x\Big)^{1/6} \Big(\int\limits_\Omega |x|^{\frac{6q(l-1-\beta)}{5q-6}}\, \mathrm{d}x\Big)^{\frac{5q-6}{6q}} \\
&\leq c\,\big\|\mathbf{f}; V^{0,q}_{\beta-l+1}(\Omega)\big\|\big\|\boldsymbol{\eta}; L^6(\Omega)\big\| < c\,\big\|\mathbf{f}; V^{0,q}_{\beta-l+1}(\Omega)\big\|\big\|\boldsymbol{\eta}; D^1_0(\Omega)\big\|.
\end{aligned}$$

Moreover, by Lemma 2.2 (i)

$$g \in V_\beta^{l,q}(\Omega) \subset V_{\beta-l-3/2+3/q}^{0,2}(\Omega) \subset L^2(\Omega).$$

Therefore, there exists a unique weak solution $(\mathbf{u}, \Pi) \in V_0^{1,2}(\Omega) \times V_0^{0,2}(\Omega)$ of problem $(3.11)_0$. As above we get the inclusion

$$\nabla\Phi \cdot \mathbf{u} \in V_{\gamma_0}^{1,2}(\Omega), \quad \Pi\nabla\Phi \in V_{\gamma_0}^{0,2}(\Omega). \tag{3.20}$$

In order to estimate the right-hand sides $\mathbf{F}$ and $G$, we use the inequality

$$\Big(\int_\Omega |h|^s|x|^{s(\mu+3/t-3/s-\varepsilon)}\,\mathrm{d}x\Big)^{1/s} \le c\Big(\int_\Omega |h|^t|x|^{\mu t}y\,\mathrm{d}x\Big)^{1/t} \tag{3.21}$$

which is true for arbitrary $h \in V_\mu^{0,t}(\Omega)$, $1 < s < t$ and $\forall\varepsilon > 0$; can be easily proved by applying Hölder's inequality.

Let us consider the case $q \in [6/5, 2]$. We take $\varepsilon_0 = \gamma_0 - 1$. Then (3.20) and (3.21) with $s = q$, $t = 2$, $\mu = \gamma_0$ imply $\nabla\Phi \cdot \mathbf{u} \in V_{1+3/2-3/q}^{1,q}(\Omega)$, $\Pi\nabla\Phi \in V_{1+3/2-3/q}^{0,q}(\Omega)$. Since $1 + 3/2 - 3/q < \beta - l + 1$, we conclude

$$\mathbf{F} \in V_{1+3/2-3/q}^{0,q}(\Omega), \quad G \in V_{1+3/2-3/q}^{1,q}(\Omega). \tag{3.22}$$

Let $q \in (2, 6]$. From (3.21) with $t = q$, $s = 2$, $\mu = \beta - l + 1$ follows that

$$\begin{aligned}(\mathbf{f}, g) &\in V_{\beta-l+1+3/q-3/2-\varepsilon_1}^{0,2}(\Omega) \times V_{\beta-l+1+3/q-3/2-\varepsilon_1}^{1,2}(\Omega)\\ &\subset V_{1-\varepsilon_1}^{0,2}(\Omega) \times V_{1-\varepsilon_1}^{1,2}(\Omega), \quad \forall\varepsilon_1 > 0.\end{aligned}$$

Since $\gamma_0 > 1$, inclusions (3.20) imply

$$\mathbf{F} \in V_{1-\varepsilon_1}^{0,2}(\Omega), \quad G \in V_{1-\varepsilon_1}^{1,2}(\Omega),$$

and, if $\varepsilon_1 < 1/2$, by Theorem 3.1 (i) we conclude

$$\mathbf{u} \in V_{1-\varepsilon_1}^{2,2}(\Omega), \quad \Pi \in V_{1-\varepsilon_1}^{1,2}(\Omega).$$

In virtue of Lemma 2.2 (i) we obtain $\mathbf{u} \in V_{-\varepsilon_1-3/q+3/2}^{1,q}(\Omega)$, $\Pi \in V_{-\varepsilon_1-3/q+3/2}^{0,q}(\Omega)$ and by Lemma 2.3 $\nabla\Phi \cdot \mathbf{u} \in V_{\gamma_0-\varepsilon_1-3/q+3/2}^{1,q}(\Omega)$, $\Pi\nabla\Phi \in V_{\gamma_0-\varepsilon_1-3/q+3/2}^{0,q}(\Omega)$. Taking $\varepsilon_1$ so that $\varepsilon_1 \le \gamma_0 - 1$, $\varepsilon_1 < 1/2$, we get

$$\nabla\Phi \cdot \mathbf{u} \in V_{1-3/q+3/2}^{1,q}(\Omega), \quad \Pi\nabla\Phi \in V_{1-3/q+3/2}^{0,q}(\Omega)$$

and, therefore, $(\mathbf{F}, G)$ satisfy inclusions (3.22).

Finally, let $q > 6$. Then inequality (3.21) gives

$$\mathbf{f} \in V_{\beta-l+1+3/q-3/2-\varepsilon_1}^{0,2}(\Omega) \subset V_{1-\varepsilon_1}^{0,2}(\Omega), \quad g \in V_{1-\varepsilon_1}^{1,2}(\Omega)$$

and as in the previous case we derive the inclusions

$$\begin{aligned}&(\mathbf{u}, \Pi) \in V_{1-\varepsilon_1}^{2,2}(\Omega) \times V_{1-\varepsilon_1}^{1,2}(\Omega) \overset{\text{Lemma 2.2}}{\Longrightarrow} (\mathbf{u}, \Pi) \in V_{1-\varepsilon_1}^{1,6}(\Omega) \times V_{1-\varepsilon_1}^{0,6}(\Omega)\\ &\overset{\text{Lemma 2.3}}{\Longrightarrow} \nabla\Phi \cdot \mathbf{u} \in V_{\gamma_0+1-\varepsilon_1}^{1,6}(\Omega), \quad \Pi\nabla\Phi \in V_{\gamma_0+1-\varepsilon_1}^{0,6}(\Omega).\end{aligned}$$

By inequality (3.21) with $\mu = \beta - l + 1,\ s = 6,\ t = q$ we have

$$\mathbf{f} \in V^{0,6}_{\beta-l+1/2+3/q-\varepsilon_1}(\Omega) \subset V^{0,6}_{2-\varepsilon_1}(\Omega), \quad g \in V^{1,6}_{2-\varepsilon_1}(\Omega).$$

Hence,

$$\mathbf{F} \in V^{0,6}_{2-\varepsilon_1}(\Omega), \quad G \in V^{1,6}_{2-\varepsilon_1}(\Omega).$$

Since $\varepsilon_1 < 1/2$, we have $(2-\varepsilon_1) - 1 \in (1/2, 3/2)$ and by Theorem 3.1 (i) we get $\mathbf{u} \in V^{2,6}_{2-\varepsilon_1}(\Omega)$, $\Pi \in V^{1,6}_{2-\varepsilon_1}(\Omega)$. Lemma 2.2 implies $\mathbf{u} \in \Lambda^{1,1/2}_{2-\varepsilon_1}(\Omega)$, $\Pi \in \Lambda^{0,1/2}_{1-\varepsilon_1}(\Omega)$. It is easy to compute that the last relations imply the inclusions

$$\nabla\Phi \cdot \mathbf{u} \in V^{1,q}_{\gamma_0+3/2-3/q-\varepsilon_1}(\Omega) \subset V^{1,q}_{1+3/2-3/q}(\Omega), \quad \Pi\nabla\Phi \in V^{0,q}_{1+3/2-3/q}(\Omega).$$

Thus, we have proved the inclusions (3.22) for arbitrary $q \in [6/5, \infty)$. Now, arguing as in the case $q = 2$, by repeated application of Theorem 3.1 (i) and Lemma 2.3 we derive the sequence of inclusions

$$\begin{aligned}&(\mathbf{F}, G) \in V^{0,q}_{\beta_0}(\Omega) \times V^{1,q}_{\beta_0}(\Omega) \overset{\text{Theorem 3.1 (i)}}{\Longrightarrow} (\mathbf{u}, \Pi) \in \mathcal{D}^{1,q}_{\beta_0}V(\Omega)\\ &\overset{\text{Lemma 2.3}}{\Longrightarrow} \dots \overset{\text{Theorem 3.1 (i)}}{\Longrightarrow} (\mathbf{u}, \Pi) \in \mathcal{D}^{l,q}_{\beta_l}V(\Omega)\\ &\overset{\text{Lemma 2.3}}{\Longrightarrow} \dots \overset{\text{Theorem 3.1 (i)}}{\Longrightarrow} (\mathbf{u}, \Pi) \in \mathcal{D}^{l,q}_{\beta_{l+m}}V(\Omega) \subset \mathcal{D}^{l,q}_{\beta}V(\Omega),\end{aligned}$$

where $\beta_0 = 1 + 3/2 - 3/q, \dots, \beta_l = \min(\gamma_0 l + \beta_0, \beta), \dots, \beta_{l+m} = \min\big((l+m)\gamma_0 + \beta_0, \beta\big)$, $(l+m)\gamma_0 + \beta_0 \geq \beta$. Supplying the obtained inclusion with the corresponding estimates we obtain estimate (3.16) for arbitrary $q \in [6/5, \infty)$.

(ii) Let $(\mathbf{f}, g) \in \Lambda^{l-1,\delta}_{\beta}(\Omega) \times \Lambda^{l,\delta}_{\beta}(\Omega)$, $l \geq 1$, $\delta \in (0,1)$, $\beta \in (l+\delta+3/2, l+\delta+2)$. Let us take $q = \frac{3}{1-\delta}$. It is easy to compute that then $(\mathbf{f}, g) \in V^{0,q}_{5/2-3/q+\widehat{\varepsilon}_0}(\Omega) \times V^{1,q}_{5/2-3/q+\widehat{\varepsilon}_0}(\Omega)$ with $2\widehat{\varepsilon}_0 = \beta - l - \delta - 3/2 > 0$. It is already proved that problem $(3.11)_0$ has a unique solution $(\mathbf{u}, \Pi) \in V^{2,q}_{5/2-3/q+\widehat{\varepsilon}_0}(\Omega) \times V^{1,q}_{5/2-3/q+\widehat{\varepsilon}_0}(\Omega)$ and by Lemma 2.2 we get the inclusion $(\mathbf{u}, \Pi) \in \Lambda^{1,\delta}_{\widehat{\beta}_0}(\Omega) \times \Lambda^{0,\delta}_{\widehat{\beta}_0}(\Omega)$, where $\widehat{\beta}_0 = 3/2 + \delta + \widehat{\varepsilon}_0$. Now, by Lemma 2.3

$$\nabla\Phi \cdot \mathbf{u} \in \Lambda^{1,\delta}_{\widehat{\beta}_0+\gamma_0}(\Omega), \quad \Pi\nabla\Phi \in \Lambda^{0,\delta}_{\widehat{\beta}_0+\gamma_0}(\Omega).$$

Therefore, $(\mathbf{F}, G) \in \Lambda^{0,\delta}_{\widehat{\beta}_1}(\Omega) \times \Lambda^{1,\delta}_{\widehat{\beta}_1}(\Omega)$, where $\widehat{\beta}_1 = \min\{\beta - l + 1, \widehat{\beta}_0 + \gamma_0\}$. In virtue of Theorem 3.1 (ii) we get

$$(\mathbf{u}, \Pi) \in \Lambda^{2,\delta}_{\widehat{\beta}_1}(\Omega) \times \Lambda^{1,\delta}_{\widehat{\beta}_1}(\Omega) = \mathcal{D}^{1,\delta}_{\widehat{\beta}_1}\Lambda(\Omega).$$

Repeating the last argument we derive the sequence of inclusions

$$(\mathbf{u},\Pi)\in\mathcal{D}^{1,\delta}_{\widehat{\beta}_1}\Lambda(\Omega)$$

$$\overset{\text{Lemma 2.3}}{\Longrightarrow}(\mathbf{F},G)\in\Lambda^{1,\delta}_{\widehat{\beta}_2}(\Omega)\times\Lambda^{2,\delta}_{\widehat{\beta}_2}(\Omega),\widehat{\beta}_2=\min\{\beta-l+2,\,\widehat{\beta}_0+2\gamma_0\}$$

$$\Longrightarrow(\mathbf{u},\Pi)\in\mathcal{D}^{2,\delta}_{\widehat{\beta}_2}\Lambda(\Omega)$$

$$\Longrightarrow\dots\overset{\text{Lemma 2.3}}{\Longrightarrow}(\mathbf{F},G)\in\Lambda^{l-1,\delta}_{\widehat{\beta}_{l+m}}(\Omega)\times\Lambda^{l,\delta}_{\widehat{\beta}_{l+m}}(\Omega)$$

$$\overset{\text{Theorem 3.1 (ii)}}{\Longrightarrow}(\mathbf{u},\Pi)\in\mathcal{D}^{l,\delta}_{\widehat{\beta}_{l+m}}\Lambda(\Omega),\quad\widehat{\beta}_{l+m}=\min\big\{\beta,\widehat{\beta}_0+(l+m)\gamma_0\big\}.$$

Taking $m$ so, that $\widehat{\beta}_0+(l+m)\gamma_0\geq\beta$, we obtain $(\mathbf{u},\Pi)\in\mathcal{D}^{l,\delta}_{\beta}\Lambda(\Omega)$. The theorem is proved. □

Let us consider problem (3.11) with nonhomogeneous boundary condition.

**Theorem 3.8.** (i) *Let* $(\mathbf{f},g,\mathbf{h})\in\mathcal{R}^{l,q}_{\beta}V(\Omega)$, $l\geq1$, $q\geq6/5$, $\beta\in(l+3/2-3/q,\,l+2-3/q)$. *Then problem* (3.11) *has a unique solution* $(\mathbf{u},\Pi)\in\mathcal{D}^{l,q}_{\beta}V(\Omega)$ *and there holds the estimate*

$$\big\|(\mathbf{u},\Pi);\mathcal{D}^{l,q}_{\beta}V(\Omega)\big\|\leq c\,\big\|(\mathbf{f},g,\mathbf{h});\mathcal{R}^{l,q}_{\beta}V(\Omega)\big\|.\tag{3.23}$$

(ii) *Let* $(\mathbf{f},g,\mathbf{h})\in\mathcal{R}^{l,\delta}_{\beta}\Lambda(\Omega)$, $l\geq1$, $\delta\in(0,1)$, $\beta\in(l+\delta+3/2,\,l+\delta+2)$. *Then problem* (3.11) *has a unique solution* $(\mathbf{u},\Pi)\in\mathcal{D}^{l,\delta}_{\beta}\Lambda(\Omega)$ *and*

$$\big\|(\mathbf{u},\Pi);\mathcal{D}^{l,\delta}_{\beta}\Lambda(\Omega)\big\|\leq c\,\big\|(\mathbf{f},g,\mathbf{h});\mathcal{R}^{l,\delta}_{\beta}\Lambda(\Omega)\big\|.\tag{3.24}$$

*Proof.* Since $\mathbf{h}\in W^{l+1-1/q,q}(\partial\Omega)$ $\big(\mathbf{h}\in C^{l+1,\delta}(\partial\Omega)\big)$, it can be extended as a function $\mathbf{H}$ belonging to $W^{l+1,q}(\Omega)$ $\big(C^{l+1,\delta}(\Omega)\big)$ and having a compact support. Hence, $\mathbf{H}\in V^{l+1,q}_{\beta}(\Omega)$ $\big(\mathbf{h}\in\Lambda^{l+1,\delta}_{\beta}(\Omega)\big)$. For $\mathbf{U}=\mathbf{u}-\mathbf{H}$ and $\Pi$ we get problem $(3.11)_0$ with the new right-hand side $(\mathbf{f}_1,g_1,\mathbf{0})$. Theorem 3.6 follows now from Theorem 3.5. □

Now we study problem (3.11) in weighted functions spaces with detached asymptotics.

**Theorem 3.9.** (i) *Let* $(\mathbf{f},g,\mathbf{h})\in\mathcal{R}^{l,q}_{\gamma}\mathfrak{V}(\Omega)$, $l\geq1$, $q\in[6/5,\infty)$, $\gamma\in(l+2-3/q,\,l+3-3/q)$ *and let* $\mathfrak{F}(\theta)$ *satisfies the compatibility condition* (3.8). *Then problem* (3.11) *has a unique solution* $(\mathbf{u},\Pi)\in\mathcal{D}^{l,q}_{\gamma}\mathfrak{V}(\Omega)$. *There holds the estimate*

$$\big\|(\mathbf{u},\Pi);\mathcal{D}^{l,q}_{\gamma_1}\mathfrak{V}(\Omega)\big\|\leq c\,\big\|\mathbf{f},g,\mathbf{h});\mathcal{R}^{l,q}_{\gamma}\mathfrak{V}(\Omega)\big\|.\tag{3.25}$$

(ii) *Let* $(\mathbf{f},g,\mathbf{h})\in\mathcal{R}^{l,\delta}_{\gamma}\mathfrak{C}(\Omega)$, $l\geq1$, $\delta\in(0,1)$, $\gamma\in(l+2+\delta,\,l+3+\delta)$ *and let the compatibility condition* (3.8) *be valid. Then problem* (3.11) *has a unique solution* $(\mathbf{u},\Pi)\in\mathcal{D}^{l,\delta}_{\gamma}\mathfrak{C}(\Omega)$ *and*

$$\big\|(\mathbf{u},\Pi);\mathcal{D}^{l,\delta}_{\gamma}\mathfrak{C}(\Omega)\big\|\leq c\,\big\|(\mathbf{f},g,\mathbf{h});\mathcal{R}^{l,\delta}_{\gamma}\mathfrak{C}(\Omega)\big\|.\tag{3.26}$$

*Proof.* (i) Since $\mathcal{R}^{l,q}_\gamma\mathfrak{V}(\Omega) \subset \mathcal{R}^{l,q}_\beta V(\Omega)$ with arbitrary $\beta \in (l+3/2-3/q,\, l+2-3/q)$, by Theorem 3.5 there exists a unique solution $(\mathbf{u}, \Pi) \in \mathcal{D}^{l,q}_{\beta_*}V(\Omega)$, $\beta_* = l+2-3/q-\varepsilon$, $0<\varepsilon<\min(1/2,\, \gamma_0-1)$. From Lemma 2.3 it follows that

$$\nabla\Phi\cdot\mathbf{u} \in V^{l+1,q}_{\beta_*+\gamma_0}(\Omega) \subset V^{l+1,q}_\gamma(\Omega), \quad \Pi\nabla\Phi \in V^{l,q}_{\beta_*+\gamma_0}(\Omega) \subset V^{l,q}_\gamma(\Omega).$$

Therefore, $(\mathbf{u}, \Pi)$ can be considered as a solution of Stokes problem (3.1) with the right-hand side $(\mathbf{F}, G, \mathbf{h}) \in \mathcal{R}^{l,q}_\gamma\mathfrak{V}(\Omega)$. In virtue of Theorem 3.3 we conclude that $(\mathbf{u}, \Pi) \in \mathcal{D}^{l,q}_\gamma\mathfrak{V}(\Omega)$ and the estimate (3.25) holds true.

(ii) The proof of the second part of the theorem is completely analogous to the that of the first part. One have just to change weighted Sobolev spaces into weighted Hölder spaces and to use Theorem 3.4 instead of Theorem 3.3. □

## 4. Transport equation and Poisson-type equation

### 4.1. Transport equation

Let us consider the transport equation

$$z + \operatorname{div}(\mathbf{w}z) = h, \quad x \in \Omega, \tag{4.1}$$

where $\mathbf{w}$ satisfies the condition

$$\mathbf{w}\cdot\mathbf{n} = 0, \quad x \in \partial\Omega \tag{4.2}$$

Problem (4.1), (4.2) was studied in many papers using various functional settings (e.g [12, 13] and references cited there). Here we need results concerning the solvability of (4.1), (4.2) in weighted Sobolev and Hölder spaces with detached asymptotics.

**Theorem 4.1.** (i) *Let* $h \in \mathfrak{V}^{(l,l+1),q}_{\gamma,2}(\Omega)$ *with* $l \ge 2$, $q > 3/2$,

$$\gamma \in (l+2-3/q,\, l+3-3/q) \tag{4.3}$$

*and let* $\mathbf{w} \in \mathfrak{V}^{(l+1,l+2),q}_{\gamma,1}(\Omega)$. *There exists a number* $\varepsilon_0 > 0$ *such that if*

$$\big\|\mathbf{w}; \mathfrak{V}^{(l+1,l+2),q}_{\gamma,1}(\Omega)\big\| \le \varepsilon_0, \tag{4.4}$$

*then problem* (4.1), (4.2) *has just one solution* $z$ *with*

$$z \in \mathfrak{V}^{(l,l+1),q}_{\gamma,2}(\Omega),\ \operatorname{div}(\mathbf{w}z) \in V^{l,q}_\gamma(\Omega). \tag{4.5}$$

*There holds the estimate*

$$\big\|z; \mathfrak{V}^{(l,l+1),q}_{\gamma,2}(\Omega)\big\| + \big\|\operatorname{div}(\mathbf{w}z); V^{l,q}_\gamma(\Omega)\big\| \le c\big\|h; \mathfrak{V}^{(l,l+1),q}_{\gamma,2}(\Omega)\big\|. \tag{4.6}$$

*Furthermore,* $\Delta z \in \mathfrak{V}^{(l-2,l-1),q}_{\gamma,4}(\Omega)$, $\Delta\operatorname{div}(\mathbf{w}z) \in V^{l-2,q}_\gamma(\Omega)$ *and*

$$\begin{aligned}&\big\|\Delta z; \mathfrak{V}^{(l-2,l-1),q}_{\gamma,4}(\Omega)\big\| + \big\|\Delta\operatorname{div}(\mathbf{w}z); V^{l-2,q}_\gamma(\Omega)\big\| \\ &\quad\le c\Big(\big\|h; \mathfrak{V}^{(l,l+1),q}_{\gamma,2}(\Omega)\big\| + \big\|\mathbf{w}; \mathfrak{V}^{(l+1,l+2),q}_{\gamma,1}(\Omega)\big\|\big\|z; \mathfrak{V}^{(l,l+1),q}_{\gamma,2}(\Omega)\big\|\Big).\end{aligned} \tag{4.7}$$

(ii) *Let* $h \in \mathfrak{C}^{(l,l+1),\delta}_{\gamma,2}(\Omega)$, $\quad l \geq 2, \quad \delta \in (0,1)$,

$$\gamma \in (l+2+\delta,\, l+3+\delta) \tag{4.8}$$

*and let* $\mathbf{w} \in \mathfrak{C}^{(l+1,l+2),\delta}_{\gamma,1}(\Omega)$. *There exists a number* $\varepsilon_0 > 0$ *such that if*

$$\big\|\mathbf{w}; \mathfrak{C}^{(l+1,l+2),\delta}_{\gamma,1}(\Omega)\big\| \leq \varepsilon_0,$$

*then problem* (4.1), (4.2) *has just one solution* $z$ *with* $z \in \mathfrak{C}^{(l,l+1),\delta}_{\gamma,2}(\Omega)$, $\operatorname{div}(\mathbf{w}z) \in \Lambda^{l,\delta}_{\gamma}(\Omega)$ *and*

$$\big\|z; \mathfrak{C}^{(l,l+1),\delta}_{\gamma,2}(\Omega)\big\| + \big\|\operatorname{div}(\mathbf{w}z); \Lambda^{l,\delta}_{\gamma}(\Omega)\big\| \leq c\,\big\|h; \mathfrak{C}^{(l,l+1),\delta}_{\gamma,2}(\Omega)\big\|. \tag{4.9}$$

*Moreover,* $\Delta z \in \mathfrak{C}^{(l-2,l-1),\delta}_{\gamma,4}(\Omega)$, $\Delta \operatorname{div}(\mathbf{w}z) \in \Lambda^{l-2,\delta}_{\gamma}(\Omega)$,

$$\begin{aligned}&\big\|\Delta z; \mathfrak{C}^{(l-2,l-1),\delta}_{\gamma,4}(\Omega)\big\| + \big\|\Delta \operatorname{div}(\mathbf{w}z); \Lambda^{l-2,\delta}_{\gamma}(\Omega)\big\| \\ &\quad \leq c\Big(\big\|h; \mathfrak{C}^{(l,l+1),\delta}_{\gamma,2}(\Omega)\big\| + \big\|\mathbf{w}; \mathfrak{C}^{(l+1,l+2),\delta}_{\gamma,1}(\Omega)\big\|\big\|z; \mathfrak{C}^{(l,l+1),\delta}_{\gamma,2}(\Omega)\big\|\Big).\end{aligned} \tag{4.10}$$

*Proof.* (i) Let $h \in \mathfrak{V}^{(l,l+1),q}_{\gamma,2}(\Omega)$, i.e. $h$ admits the representation

$$h(x) = r^{-2}\mathfrak{H}(\theta) + \widetilde{h}(x).$$

with attributes $\mathfrak{H} \in W^{l+1,q}(\mathbb{S})^2$ and $\widetilde{h} \in V^{l,q}_{\gamma}(\Omega)$. For sufficiently small $\varepsilon_0$ in [10] is proved the existence of the unique solution $z \in \mathfrak{V}^{(l,l+1),q}_{\gamma,2}(\Omega)$ of problem (4.1), (4.2) which admits the asymptotic representation

$$z(x) = r^{-2}\mathfrak{H}(\theta) + \widetilde{z}(x), \quad \widetilde{z} \in V^{l,q}_{\gamma}(\Omega) \tag{4.11}$$

i.e. the "spherical" attributes of $h$ and $z$ coincide. Moreover, there holds the estimate

$$\big\|\widetilde{z}; V^{l,q}_{\gamma}(\Omega)\big\| \leq c\,\big\|h; \mathfrak{V}^{(l,l+1),q}_{\gamma,2}(\Omega)\big\| \tag{4.12}$$

Since $\widetilde{z}$ and $\widetilde{h}$ have the same "spherical" attributes, from equation (4.1) we get $\operatorname{div}(\mathbf{w}z) = \widetilde{h} - \widetilde{z} \in V^{l,q}_{\gamma}(\Omega)$ and estimate (4.6) follows from (4.12).

Applying the operator $\Delta$ to equation (4.1) we obtain

$$\Delta z + \operatorname{div}(\mathbf{w}\Delta z) = \Delta h - \operatorname{div}(\Delta \mathbf{w} z) - 2\operatorname{div}(\nabla \mathbf{w} \cdot \nabla z). \tag{4.13}$$

From Lemmata 2.1 and 2.5 we conclude

$$\Delta h \in \mathfrak{V}^{(l-2,l-1)}_{\gamma,4}(\Omega),\ \operatorname{div}(\Delta \mathbf{w} z) + 2\operatorname{div}(\nabla \mathbf{w} \cdot \nabla z) \in V^{l-2,q}_{\gamma}(\Omega)$$

and

$$\begin{aligned}&\big\|\operatorname{div}(\Delta \mathbf{w} z) + 2\operatorname{div}(\nabla \mathbf{w} \cdot \nabla z); V^{l-2,q}_{\gamma}(\Omega)\big\| \\ &\quad \leq c\,\big\|\mathbf{w}; \mathfrak{V}^{(l+1,l+2),q}_{\gamma,1}(\Omega)\big\|\big\|z; \mathfrak{V}^{(l,l+1),q}_{\gamma,2}(\Omega)\big\|.\end{aligned} \tag{4.14}$$

Therefore, $\Delta z \in \mathfrak{V}^{(l-2,l-1),q}_{\gamma,4}(\Omega)$ may be interpreted as a unique solution of (4.1) with the right-hand side $h_1 = \Delta h - \operatorname{div}(\Delta \mathbf{w} z) - 2\operatorname{div}(\nabla \mathbf{w} \cdot \nabla z) \in \mathfrak{V}^{(l-2,l-1),q}_{\gamma,4}(\Omega)$. Now, from results of the paper [10] (see Section 4) follows the estimate

$$\big\|\Delta z; \mathfrak{V}^{(l-2,l-1),q}_{\gamma,4}(\Omega)\big\| \leq c\,\big\|h_1; \mathfrak{V}^{(l-2,l-1),q}_{\gamma,4}(\Omega)\big\|. \tag{4.15}$$

Since div $(\mathbf{w}\Delta z) = h_1 - \Delta z$, from (4.14), (4.15) we derive estimate (4.7).

(ii) The proof is completely analogous to that of the case (i). We only note that the existence of the unique solution $z \in \mathfrak{C}^{(l,l+1),\delta}_{\gamma,2}(\Omega)$ to (4.1) is proved in [5]. □

*Remark* 4.2. Representing the solution $z$ in the form (4.11) we get for the remainder $\widetilde{z}$ the transport equation (4.1) with the right-hand side $h - \operatorname{div}\big(\mathbf{w}r^{-2}\mathfrak{H}(\theta)\big)$. That oblige us to require more regularity for the "spherical" attributes of $z$ and $h$. This choise is possible only because they coinside.

**4.2. Boundary value problems for the Poisson equation**

Let us consider Neumann problem for the operator $\Delta_{\rho_0} = \operatorname{div}\big(\rho_0(x)\nabla\big)$, where $\rho_0(x) = \rho_* \exp\Phi(x)$, $\Phi \in \Lambda^{l+1,\delta}_{l+1+\delta+\gamma_0}(\Omega)$, $\gamma_0 > 1$, i.e. we consider the problem

$$\begin{cases} -\Delta_{\rho_0}\varphi = \psi, & x \in \Omega, \\ \dfrac{\partial\varphi}{\partial n} = 0, & x \in \partial\Omega, \quad \varphi(x) \to 0, \ |x| \to \infty. \end{cases} \tag{4.16}$$

**Theorem 4.3.** (i) *Let* $\psi \in V^{l-1,q}_\gamma(\Omega)$, $l \ge 1$, $q > 1$, $\gamma \in (l+2-3/q,\, l+3-3/q)$. *Then problem* (4.16) *has a unique solution* $\varphi$ *which admits the asymptotic representation*

$$\varphi(x) = c_0\big(2\pi|x|\big)^{-1} + \widetilde{\varphi}(x), \tag{4.17}$$

*where* $\widetilde{\varphi} \in V^{l+1,q}_\gamma(\Omega)$. *There holds the estimate*

$$\big\|\widetilde{\varphi}; V^{l+1,q}_\gamma(\Omega)\big\| + |c_0| \le c\,\big\|\psi; V^{l-1,q}_\gamma(\Omega)\big\|. \tag{4.18}$$

(ii) *Let* $\psi \in \Lambda^{l-1,\delta}_\gamma(\Omega)$, $l \ge 1$, $\delta \in (0,1)$, $\gamma \in (l+2+\delta,\, l+3+\delta)$. *Then problem* (4.16) *has a unique solution* $\varphi$ *which admits the asymptotic representation* (4.17) *with* $\widetilde{\varphi} \in \Lambda^{l+1,\delta}_\gamma(\Omega)$ *and there holds the estimate*

$$\big\|\widetilde{\varphi}; \Lambda^{l+1,\delta}_\gamma(\Omega)\big\| + |c_0| \le c\,\big\|\psi; \Lambda^{l-1,\delta}_\gamma(\Omega)\big\|. \tag{4.19}$$

The proof of Theorem 4.3 can be found in [9].

Let us consider now the Dirichlet problem for the Laplace operator

$$\begin{cases} -\Delta\xi = \zeta, & x \in \Omega, \\ \xi = 0, & x \in \partial\Omega. \end{cases} \tag{4.20}$$

**Theorem 4.4.** [9]. *Let* $\zeta \in V^{l-1,q}_\gamma(\Omega)$, $l \ge 1$, $q > 1$, $\gamma > l+1-3/q$ *and let* $\xi \in V^{0,q}_{\gamma-l-1}(\Omega)$ *be a solution of problem* (4.20). *Then* $\xi \in V^{l+1,q}_\gamma(\Omega)$ *and*

$$\big\|\xi; V^{l+1,q}_\gamma(\Omega)\big\| \le c\,\big\|\zeta; V^{l-1,q}_\gamma(\Omega)\big\|. \tag{4.21}$$

*The solution* $\xi \in V^{0,q}_{\gamma-l-1}(\Omega)$ *is unique.*

## 5. Linearized problem

Let us consider problem (1.9′).

**Theorem 5.1.** *Let* $\Omega \in \mathbb{R}^3$ *be an exterior domain with the smooth boundary* $\partial\Omega$, $\Phi \in \Lambda^{l+1,\delta}_{l+1+\delta+\gamma_0}(\Omega)$, $\gamma_0 > 1$, $l \geq 2$, $\delta \in (0,1)$, $\rho_0(x) = \rho_* \exp \Phi(x)$. *Suppose that* $\mathbf{w} \in \mathfrak{V}^{(l+1,l+2),q}_{\gamma,1}(\Omega)$, $\mathbf{F} \in \mathfrak{V}^{(l-1,l),q}_{\gamma,3}(\Omega)$, $g \in V^{l,q}_{\gamma}(\Omega)$, $q > 3/2$, $\gamma \in (l+2-3/q,\, l+3-3/q)$, *be given functions with*

$$\mathbf{w} = 0, \quad \operatorname{div} \mathbf{w} = 0, \quad x \in \partial\Omega, \tag{5.1}$$

*and* $\mathbf{F}$ *satisfies the compatibility condition* (3.8). *There exists a number* $\varepsilon_0 > 0$ *such that if*

$$\left\|\mathbf{w}; \mathfrak{V}^{(l+1,l+2),q}_{\gamma,1}(\Omega)\right\| < \varepsilon_0, \tag{5.2}$$

*then problem* (1.9′) *has a unique solution* $(\sigma, \mathbf{v}) \in \mathfrak{V}^{(l,l+1),q}_{\gamma,2}(\Omega) \times \mathfrak{V}^{(l+1,l+2),q}_{\gamma,1}(\Omega)$. *Moreover,* $\operatorname{div}(\mathbf{w}\sigma) \in V^{l+1,q}_{\gamma}(\Omega)$ *and there hold the estimates*

$$\begin{aligned} &\left\|\sigma; \mathfrak{V}^{(l,l+1),q}_{\gamma,2}(\Omega)\right\| + \left\|\mathbf{v}; \mathfrak{V}^{(l+1,l+2),q}_{\gamma,1}(\Omega)\right\| \\ &\qquad \leq c\left(\left\|\mathbf{F}; \mathfrak{V}^{(l-1,l)}_{\gamma,3}, q(\Omega)\right\| + \left\|g; V^{l,q}_{\gamma}(\Omega)\right\|\right), \end{aligned} \tag{5.3}$$

$$\left\|\operatorname{div}(\mathbf{w}\sigma); V^{l,q}_{\gamma}(\Omega)\right\| \leq c\left(\left\|\mathbf{F}; \mathfrak{V}^{(l-1,l),q}_{\gamma,3}(\Omega)\right\| + \left\|g; V^{l,q}_{\gamma}(\Omega)\right\|\right). \tag{5.4}$$

*Proof.* As it is mentioned in Introduction, problem (1.9′) is equivalent to three problems (1.13)–(1.15) and the solution $(\sigma, \mathbf{v})$ of (1.9′) can be found as a fixed point of the linear mapping $\mathcal{L}\colon \tau \to \sigma$ defined by (1.13)-(1.15). Let us give exact meaning to the formal decomposition scheme described in Introduction. Define the Banach space

$$\mathfrak{B}^{l,q}_{\gamma}(\Omega) = \left\{\sigma\colon \sigma \in \mathfrak{V}^{(l,l+1),q}_{\gamma,2}(\Omega),\ \Delta\operatorname{div}(\mathbf{w}\sigma) \in V^{l-2,q}_{\gamma}(\Omega)\right\}$$

with the norm

$$\left\|\sigma; \mathfrak{B}^{l,q}_{\gamma}(\Omega)\right\| = \left\|\sigma; \mathfrak{V}^{(l,l+1),q}_{\gamma,2}(\Omega)\right\| + \left\|\Delta\big(\operatorname{div}(\mathbf{w}\sigma)\big); V^{l-2,q}_{\gamma}(\Omega)\right\|.$$

Let $\tau \in \mathfrak{B}^{l,q}_{\gamma}(\Omega)$ be given. First, we find $\varphi$ as a solution of the Neumann problem

$$\Delta_{\rho_0}\varphi = -\operatorname{div}(\tau\mathbf{w}) + g, \ x \in \Omega; \quad \frac{\partial\varphi}{\partial n} = 0, \ x \in \partial\Omega. \tag{5.5}$$

Second, we find the solution $(\mathbf{u}, \Pi)$ of the Stoke-type problem (1.14) taking in (1.16) $\sigma = \tau$, and, finally, we find $\sigma$ as the solution of transport equation (1.15). In such a way we define the operator $\mathcal{L}\colon \tau \to \sigma$. We show that $\mathcal{L}$ is a contraction in the space $\mathfrak{B}^{l,q}_{\gamma}(\Omega)$. The fixed point $\sigma$ of $\mathcal{L}$ and the corresponding to $\sigma$ solutions $\varphi$ (solution of (1.13)) and $(\mathbf{u}, \Pi)$ (solution of (1.14)) solve (1.13)–(1.15). The solution of the original linearized problem (1.9) has the form $(\sigma, \mathbf{v}) = (\sigma, \mathbf{u} + \nabla\varphi)$.

Consider problem (5.5). First, we realize that

$$\left\|\operatorname{div}(\mathbf{w}\tau); V^{l-1,q}_{\gamma-1}(\Omega)\right\| \leq c\left\|\tau; \mathfrak{V}^{(l,l+1),q}_{\gamma,2}(\Omega)\right\| \left\|\mathbf{w}; \mathfrak{V}^{(l+1,l+2),q}_{\gamma,1}(\Omega)\right\| \tag{5.6}$$

(see Lemma 2.5. Further, since $(\operatorname{div}\mathbf{w})|_{\partial\Omega}=0$, $\mathbf{w}|_{\partial\Omega}=0$, we get

$$\operatorname{div}(\mathbf{w}\tau)=0, \quad x\in\partial\Omega.$$

Therefore, $\operatorname{div}(\mathbf{w}\tau)$ can be considered as the solution of the Dirichlet problem

$$\begin{cases}\Delta\big(\operatorname{div}(\mathbf{w}\tau)\big)=\Delta\big(\operatorname{div}(\mathbf{w}\tau)\big), & x\in\Omega,\\ \operatorname{div}(\mathbf{w}\tau)=0, \quad x\in\partial\Omega.\end{cases} \tag{5.7}$$

By Theorem 4.4 the solution $\operatorname{div}(\mathbf{w}\tau)$ of the problem (5.7) satisfies the estimate

$$\big\|\operatorname{div}(\mathbf{w}\tau);V_\gamma^{l,q}(\Omega)\big\|\le c\Big\|\Delta\big(\operatorname{div}(\mathbf{w}\tau)\big);V_\gamma^{l-2,q}(\Omega)\Big\| \tag{5.8}$$

Now, in virtue of Theorem 4.3 there exists a unique solution $\varphi$ of problem (5.5) admiting the asymptotic representation (4.18) with $\widetilde{\varphi}\in V_\gamma^{l+2,q}(\Omega)$. Moreover, estimates (4.19), (5.6), (5.8) give

$$\begin{aligned}&\big\|\widetilde{\varphi};V_{\gamma-1}^{l+1,q}(\Omega)\big\|+|c_0|\\&\qquad\le c\Big(\big\|\tau;\mathfrak{V}_{\gamma,2}^{(l,l+1),q}(\Omega)\big\|\big\|\mathbf{w};\mathfrak{V}_{\gamma,1}^{(l+1,l+2),q}(\Omega)\big\|+\big\|g;V_{\gamma-1}^{l-1,q}(\Omega)\big\|\Big),\end{aligned} \tag{5.9}$$

$$\begin{aligned}&\big\|\widetilde{\varphi};V_{\gamma}^{l+2,q}(\Omega)\big\|+|c_0|\\&\qquad\le c\Big(\big\|\Delta\operatorname{div}(\mathbf{w}\tau);V_\gamma^{l-2,q}(\Omega)\big\|+\big\|g;V_\gamma^{l,q}(\Omega)\big\|\Big).\end{aligned} \tag{5.10}$$

Next, we consider problem (1.14) with $\mathbf{G}$ defined by (1.16) at $\sigma=\tau$. Using Lemmata 2.1–2.5 and the asymptotic representation (4.18) for $\varphi$ we conclude that $\mathbf{G}\in\mathfrak{V}_{\gamma,3}^{l-1,q}(\Omega)$ and

$$\begin{aligned}\big\|\mathbf{G};\mathfrak{V}_{\gamma,3}^{(l-1,l),q}(\Omega)\big\|&\le c\Big(\big\|\mathbf{F};\mathfrak{V}_{\gamma,3}^{(l-1,l),q}(\Omega)\big\|+\big\|\widetilde{\varphi};V_\gamma^{l+1,q}(\Omega)\big\|+|c_0|\\&+\big\|\tau;\mathfrak{V}_{\gamma,2}^{(l,l+1),q}(\Omega)\big\|\big\|\mathbf{w};\mathfrak{V}_{\gamma,1}^{(l+1,l+2),q}(\Omega)\big\|+\big\|g;V_\gamma^{l,q}(\Omega)\big\|\Big).\end{aligned} \tag{5.11}$$

Moreover, $\mathbf{G}$ and $\mathbf{F}$ have the same spherical attributes and, therefore, $\mathbf{G}$ satisfies the compatibility condition (3.8). Hence, in virtue of Theorem 3.7 (i) there exists a unique solution $(\mathbf{u},\Pi)\in\mathfrak{V}_{\gamma,1}^{(l+1,l+2)q}(\Omega)\times\mathfrak{V}_{\gamma,2}^{(l,l+l)q}(\Omega)$ of problem (1.16) and there hold the estimates

$$\begin{aligned}&\big\|\mathbf{u};\mathfrak{V}_{\gamma,1}^{(l+1,l+2),q}(\Omega)\big\|+\big\|\Pi;\mathfrak{V}_{\gamma,2}^{(l,l+1),q}(\Omega)\big\|\\&\quad\le c\Big(\big\|\mathbf{G};\mathfrak{V}_{\gamma,3}^{(l-1,l),q}(\Omega)\big\|+\big\|\nabla\widetilde{\varphi};W^{l+1-1/q,q}(\partial\Omega)\big\|+|c_0|+\big\|g;V_\gamma^{l,q}(\Omega)\big\|\Big),\end{aligned} \tag{5.12}$$

$$\begin{aligned}&\big\|\mathbf{u};\mathfrak{V}_{\gamma-1,1}^{(l,l+1),q}(\Omega)\big\|+\big\|\Pi;\mathfrak{V}_{\gamma-1,2}^{(l-1,l),q}(\Omega)\big\|\\&\quad\le c\Big(\big\|\mathbf{G};\mathfrak{V}_{\gamma-1,3}^{(l-2,l-1),q}(\Omega)\big\|+\big\|\nabla\widetilde{\varphi};W^{l-1/q,q}(\partial\Omega)\big\|+|c_0|+\big\|g;V_{\gamma-1}^{l-1,q}(\Omega)\big\|\Big).\end{aligned} \tag{5.13}$$

Using estimates (5.9)–(5.11) and the trace theorem, from (5.12), (5.13) we obtain

$$\begin{aligned}
&\|\mathbf{u};\mathfrak{V}^{(l+1,l+2),q}_{\gamma,1}(\Omega)\| + \|\Pi;\mathfrak{V}^{(l,l+1),q}_{\gamma,2}(\Omega)\| \\
&\le c_* \Big(\|\mathbf{F};\mathfrak{V}^{(l-1,l),q}_{\gamma,3}(\Omega)\| + \|\tau;\mathfrak{V}^{(l,l+1),q}_{\gamma,2}(\Omega)\|\|\mathbf{w};\mathfrak{V}^{(l+1,l+2),q}_{\gamma,1}(\Omega)\| \\
&\quad + \|\Delta \mathrm{div}\,(\mathbf{w}\tau);V^{l-2,q}_{\gamma}(\Omega)\| + \|g;V^{l,q}_{\gamma}(\Omega)\|\Big)
\end{aligned} \tag{5.14}$$

and

$$\begin{aligned}
&\|\mathbf{u};\mathfrak{V}^{(l,l+1),q}_{\gamma-1,1}(\Omega)\| + \|\Pi;\mathfrak{V}^{(l-1,l),q}_{\gamma-1,2}(\Omega)\| \le c\Big(\|\mathbf{F};\mathfrak{V}^{(l-2,l-1),q}_{\gamma-1,3}(\Omega)\| \\
&\quad + \|\tau;\mathfrak{V}^{(l,l+1),q}_{\gamma,2}(\Omega)\|\|\mathbf{w};\mathfrak{V}^{(l+1,l+2),q}_{\gamma,1}(\Omega)\| + \|g;V^{l-1,q}_{\gamma-1}(\Omega)\|\Big).
\end{aligned} \tag{5.15}$$

Moreover, applying the operator div to equation $(1.14)_1$ and using the relation

$$\mathrm{div}\,\mathbf{u} = -\rho_0^{-1}\nabla\rho_0\cdot\mathbf{u},$$

we get

$$\Delta\Pi = \mathrm{div}\,\mathbf{G} + \mathrm{div}\,(\rho_0^{-1}\nabla\rho_0\Pi) - (2\mu_1+\mu_2)\Delta(\rho_0^{-1}\nabla\rho_0\cdot\mathbf{u}).$$

Therefore, in virtue of (5.15), (5.11), (5.9),

$$\begin{aligned}
&\|\Delta\Pi;\mathfrak{V}^{(l-2,l-1),q}_{\gamma,4}(\Omega)\| \\
&\le c\Big(\|\mathbf{F};\mathfrak{V}^{(l-1,l),q}_{\gamma,3}(\Omega)\| + \|\tau;\mathfrak{V}^{(l,l+1),q}_{\gamma,2}(\Omega)\|\|\mathbf{w};\mathfrak{V}^{(l+1,l+2),q}_{\gamma,1}(\Omega)\| \\
&\quad + \|\mathbf{u};\mathfrak{V}^{(l,l+1),q}_{\gamma-1,1}(\Omega)\| + \|\Pi;\mathfrak{V}^{(l-1,l),q}_{\gamma-1,2}(\Omega)\| + \|g;V^{l,q}_{\gamma}(\Omega)\|\Big) \\
&\le c\Big(\|\mathbf{F};\mathfrak{V}^{(l-1,l),q}_{\gamma,3}(\Omega)\| + \|\tau;\mathfrak{V}^{(l,l+1),q}_{\gamma,2}(\Omega)\|\|\mathbf{w};\mathfrak{V}^{(l+1,l+2),q}_{\gamma,1}(\Omega)\| \\
&\quad + \|g;V^{l,q}_{\gamma}(\Omega)\|\Big).
\end{aligned} \tag{5.16}$$

Finally, we consider the transport equation (1.15). According to Theorem 4.1, there exists a unique solution $\sigma\in\mathfrak{V}^{(l,l+1),q}_{\gamma,2}(\Omega)$ of (1.15) and there holds the estimate

$$\begin{aligned}
&\|\sigma;\mathfrak{V}^{(l,l+1),q}_{\gamma,2}(\Omega)\| + \|\mathrm{div}\,(\mathbf{w}\sigma);V^{l,q}_{\gamma}(\Omega)\| \le c\|\Pi;\mathfrak{V}^{(l,l+1),q}_{\gamma,2}(\Omega)\| \\
&\le c_1\Big(\|\mathbf{F};\mathfrak{V}^{(l-1,l),q}_{\gamma,3}(\Omega)\| + \|\tau;\mathfrak{V}^{(l,l+1),q}_{\gamma,2}(\Omega)\|\|\mathbf{w};\mathfrak{V}^{(l+1,l+2),q}_{\gamma,1}(\Omega)\| \\
&\quad + \|\Delta \mathrm{div}\,(\mathbf{w}\tau);V^{l-2,q}_{\gamma}(\Omega)\| + \|g;V^{l,q}_{\gamma}(\Omega)\|\Big).
\end{aligned} \tag{5.17}$$

Further, estimate (4.7) yields

$$\begin{aligned}
&\|\Delta \mathrm{div}\,(\mathbf{w}\sigma);V^{l-2,q}_{\gamma}(\Omega)\| \\
&\le c\Big(\|\Delta\Pi;\mathfrak{V}^{(l-2,l-1),q}_{\gamma,4}(\Omega)\| + \|\tau;\mathfrak{V}^{(l,l+1),q}_{\gamma,2}(\Omega)\|\|\mathbf{w};\mathfrak{V}^{(l+1,l+2),q}_{\gamma,1}(\Omega)\|\Big) \\
&\le c_2\Big(\|\mathbf{F};\mathfrak{V}^{(l-1,l),q}_{\gamma,3}(\Omega)\| + \|\tau;\mathfrak{V}^{(l,l+1),q}_{\gamma,2}(\Omega)\|\|\mathbf{w};\mathfrak{V}^{(l+1,l+2),q}_{\gamma,1}(\Omega)\| \\
&\quad + \|g;V^{l,q}_{\gamma}(\Omega)\|\Big).
\end{aligned} \tag{5.18}$$

Estimates (5.17), (5.18) together with condition (5.2) furnish

$$\begin{aligned}&\big\|\sigma;\mathfrak{V}^{(l,l+1),q}_{\gamma,2}(\Omega)\big\|+2c_1\big\|\Delta\mathrm{div}\,(\mathbf{w}\sigma);V^{l-2,q}_{\gamma}(\Omega)\big\|\\&\quad\le c_1(1+2c_2)\Big(\big\|\mathbf{F};\mathfrak{V}^{(l-1,l),q}_{\gamma,3}(\Omega)\big\|+\big\|g;V^{l,q}_{\gamma}(\Omega)\big\|+\varepsilon_0\big\|\tau;\mathfrak{V}^{(l,l+1),q}_{\gamma,2}(\Omega)\big\|\Big)\\&\qquad+c_1\big\|\Delta\mathrm{div}\,(\mathbf{w}\tau);V^{l-2,q}_{\gamma}(\Omega)\big\|.\end{aligned}\tag{5.19}$$

Thus, for every $\tau\in\mathfrak{B}^{l,q}_{\gamma}(\Omega)$ and $\mathbf{w}$ satisfying (5.1), (5.2) we have $\mathcal{L}\tau=\sigma\in\mathfrak{B}^{l,q}_{\gamma}(\Omega)$. Put $\sigma_1=\mathcal{L}\tau_1,\ \sigma_2=\mathcal{L}\tau_2$. Due to the linearity of $\mathcal{L}$ we have

$$\begin{aligned}&\big\|\sigma_1-\sigma_2;\mathfrak{V}^{(l,l+1),q}_{\gamma,2}(\Omega)\big\|+2c_1\big\|\Delta\mathrm{div}\,\big(\mathbf{w}(\sigma_1-\sigma_2)\big);V^{l-2,q}_{\gamma}(\Omega)\big\|\\&\quad\le\frac{1}{2}\Big(2c_1(1+2c_2)\varepsilon_0\big\|\tau_1-\tau_2;\mathfrak{V}^{(l,l+1),q}_{\gamma,2}(\Omega)\big\|+2c_1\big\|\Delta\mathrm{div}\,\big(\mathbf{w}(\tau_1-\tau_2)\big);V^{l-2,q}_{\gamma}(\Omega)\big\|\Big).\end{aligned}$$

Since the norms $\big\|\sigma;\mathfrak{B}^{l,q}_{\gamma}(\Omega)\big\|$ and $\Big(\big\|\sigma;\mathfrak{V}^{(l,l+1),q}_{\gamma,2}(\Omega)\big\|+2c_1\big\|\Delta\mathrm{div}\,(\sigma\mathbf{w});V^{l-2,q}_{\gamma}(\Omega)\big\|\Big)$ are equivalent, this yields the contraction in $\mathfrak{B}^{l,q}_{\gamma}(\Omega)$ provided that

$$2c_1(1+2c_2)\varepsilon_0<1.$$

Consequently, there exists a unique fixed point $\sigma$ of $\mathcal{L}$. Obviously, $\sigma$ and corresponding $\mathbf{v}=\mathbf{u}+\nabla\varphi$ solve the original linearized problem (1.9′). Estimates (5.3), (5.4) follow from (5.10), (5.14), (5.17), (5.18) written for the fixed point. The theorem is proved. □

Analogous results are valid in the scale of weighted Hölder spaces with detached asymptotics.

**Theorem 5.2.** *Let* $\Omega$, $\Phi$ *and* $\rho_0$ *be as in Theorem* 5.1. *Suppose that* $\mathbf{w}\in\mathfrak{C}^{(l+1,l+2),\delta}_{\gamma,1}(\Omega)$, $\mathbf{F}\in\mathfrak{C}^{(l-1,l),\delta}_{\gamma,3}(\Omega)$, $g\in\Lambda^{l,\delta}_{\gamma}(\Omega)$, $\delta\in(0,1)$, $\gamma\in(l+2+\delta,\,l+3+\delta)$, *be given functions such that* $\mathbf{w}$ *satisfies boundary condition* (5.1) *and* $\mathbf{F}$ *satisfies compatibility condition* (3.8). *There exists a number* $\varepsilon_0$ *such that if*

$$\big\|\mathbf{w};\mathfrak{C}^{(l+1,l+2),\delta}_{\gamma,1}(\Omega)\big\|<\varepsilon_0,$$

*then problem* (1.9′) *has a unique solution* $(\sigma,\mathbf{v})\in\mathfrak{C}^{(l,l+1),\delta}_{\gamma,2}(\Omega)\times\mathfrak{C}^{(l+1,l+2),\delta}_{\gamma,1}(\Omega)$. *Moreover,* $\mathrm{div}\,(\mathbf{w}\sigma)\in\Lambda^{l,\delta}_{\gamma}(\Omega)$ *and there holds the estimates*

$$\begin{aligned}&\big\|\sigma;\mathfrak{C}^{(l,l+1),\delta}_{\gamma,2}(\Omega)\big\|+\big\|\mathbf{v};\mathfrak{C}^{(l+1,l+2),\delta}_{\gamma,1}(\Omega)\big\|\\&\qquad\le c\Big(\big\|\mathbf{F};\mathfrak{C}^{(l-1,l),\delta}_{\gamma,3}(\Omega)\big\|+\big\|g;\Lambda^{l,\delta}_{\gamma}(\Omega)\big\|\Big),\end{aligned}\tag{5.20}$$

$$\big\|\mathrm{div}\,(\mathbf{w}\sigma);\Lambda^{l,\delta}_{\gamma}(\Omega)\big\|\le c\Big(\big\|\mathbf{F};\mathfrak{C}^{(l-1,l),\delta}_{\gamma,3}(\Omega)\big\|+\big\|g;\Lambda^{l,\delta}_{\gamma}(\Omega)\big\|\Big).\tag{5.21}$$

## 6. Nonlinear problem

Before studying the nonlinear problem (1.7), (1.8), we show that the convective term $(\rho_0+\sigma)(\mathbf{v}\cdot\nabla)\mathbf{v}$ satisfies the compatibility condition (3.8).

**Lemma 6.1.** *Let* $\mathbf{v}\in\mathfrak{V}^{(l+1,l+2),q}_{\gamma,1}(\Omega)$, $\sigma\in\mathfrak{V}^{(l,l+1)}_{\gamma,2}(\Omega)$, $\rho_0(x)=\rho_*\exp\Phi(x)$, $\Phi\in\Lambda^{l+1,\delta}_{l+1+\delta+\gamma_0}(\Omega)$, $\gamma_0>1$, $l\geq 2$, $\delta\in(0,1)$, $q>3/2$, $\gamma\in(l+2-3/q,\,l+3-3/q)$. *If*

$$\operatorname{div}\big((\rho_0+\sigma)\mathbf{v}\big)=g \ \text{in}\ \Omega,\quad g\in V^{l-1,q}_{\gamma}(\Omega), \tag{6.1}$$

*then* $(\rho_0+\sigma)(\mathbf{v}\cdot\nabla)\mathbf{v}\in\mathfrak{V}^{(l-1,l),q}_{\gamma,3}(\Omega)$ *and there holds compatibility condition* (3.8).

*Proof.* Because of condition (6.1) the convective term can be represented in the form

$$(\rho_0+\sigma)(\mathbf{v}\cdot\nabla)\mathbf{v}=\operatorname{div}\big((\rho_0+\sigma)\mathbf{v}\otimes\mathbf{v})\big)-g.$$

Further, from the inclusion $\Phi\in\Lambda^{l+1,\delta}_{l+1+\delta+\gamma_0}(\Omega)$ it follows that

$$\rho_0(x)=\rho_*+\rho_*\big(\exp\Phi(x)-1\big)\equiv\rho_*+\widetilde{\rho}_0(x),$$

with $\widetilde{\rho}_0\in\Lambda^{l+1,\delta}_{l+1+\delta+\gamma_0}(\Omega)$. In virtue of Lemma 2.4, $\mathbf{v}\otimes\mathbf{v}\in\mathfrak{V}^{(l,l+1),q}_{\gamma,2}(\Omega)$ and, therefore, Lemmata 2.3, 2.5 imply

$$(\widetilde{\rho}_0+\sigma)(\mathbf{v}\otimes\mathbf{v})\in V^{l,q}_{\gamma}(\Omega).$$

Hence,

$$(\rho_0+\sigma)(\mathbf{v}\otimes\mathbf{v})=\rho_*(\mathbf{v}\otimes\mathbf{v})+(\widetilde{\rho}_0+\sigma)(\mathbf{v}\otimes\mathbf{v})\in\mathfrak{V}^{(l,l+1),q}_{\gamma,2}(\Omega).$$

Moreover, $(\rho_0+\sigma)(\mathbf{v}\otimes\mathbf{v})$ obviously can be represented in the form

$$(\rho_0+\sigma)(\mathbf{v}\otimes\mathbf{v})=\rho_*r^{-2}\mathfrak{V}(\theta)\otimes\mathfrak{V}(\theta)+\rho_*\widetilde{\mathbf{v}}\otimes\widetilde{\mathbf{v}}+(\widetilde{\rho}_0+\sigma)(\mathbf{v}\otimes\mathbf{v}).$$

Here $\mathfrak{V}$ are $\widetilde{\mathbf{v}}$ are "attributes" of $\mathbf{v}$. Finally, we get

$$\begin{aligned}\mathbf{h}(x)&=r^{-3}\mathfrak{H}(\theta)+\widetilde{\mathbf{h}}(x)\equiv(\rho_0+\sigma)(\mathbf{v}\cdot\nabla)\mathbf{v}\\&=\rho_*\operatorname{div}\big(r^{-2}\mathfrak{V}(\theta)\otimes\mathfrak{V}(\theta)\big)+\operatorname{div}(\rho_*\widetilde{\mathbf{v}}\otimes\widetilde{\mathbf{v}})+\operatorname{div}\big[(\widetilde{\rho}_0+\sigma)\mathbf{v}\otimes\mathbf{v}\big]-g.\end{aligned}\tag{6.2}$$

By Lemma 2.1, $\mathbf{h}\in\mathfrak{V}^{(l-1,l),q}_{\gamma,3}(\Omega)$ and, since, $r^{-3}\mathfrak{H}(\theta)=\rho_*\operatorname{div}(r^{-2}\mathfrak{V}\otimes\mathfrak{V})$, Lemma 3.4 in [10] yields the condition (3.8):

$$\int_{\mathbb{S}^2}\mathfrak{H}(\theta)\,\mathrm{d}S_\theta=0. \qquad\square$$

Now we are in a position to prove the main result of the paper.

**Theorem 6.2.** *Let* $\Omega\in\mathbb{R}^3$ *be an exterior domain with the smooth boundary* $\partial\Omega$, $\Phi\in\Lambda^{l+1,\delta}_{l+1+\delta+\gamma_0}(\Omega)$, $\mathbf{f}\in\mathfrak{V}^{(l-1,l),q}_{\gamma,3}(\Omega)$, $l\geq 2$, $\delta\in(0,1)$, $q>3/2$, $\gamma\in(l+2-3/q,\,l+3-3/q)$, $\gamma_0>1$. *Suppose that the function* $\mathbf{f}$ *satisfies compatibility condition* (3.8). *There exists a number* $\varepsilon_*>0$ *such that if*

$$\big\|\mathbf{f};\mathfrak{V}^{(l-1,l),q}_{\gamma,3}(\Omega)\big\|<\varepsilon_*, \tag{6.3}$$

*then problem* (1.7), (1.8) *has exactly one solution*

$$(\sigma, \mathbf{v}) \in \mathfrak{V}^{(l,l+1),q}_{\gamma,2}(\Omega) \times \mathfrak{V}^{(l+1,l+2),q}_{\gamma,1}(\Omega)$$

*satisfying the estimate*

$$\big\|\sigma; \mathfrak{V}^{(l,l+1),q}_{\gamma,2}(\Omega)\big\| + \big\|\mathbf{v}; \mathfrak{V}^{(l+1,l+2),q}_{\gamma,1}(\Omega)\big\| \le c\big\|\mathbf{f}; \mathfrak{V}^{(l-1,l),q}_{\gamma,3}(\Omega)\big\|. \tag{6.4}$$

*Proof.* First, we describe more precisely the nonlinear map $\mathcal{N}$ defined by (1.17). Let us introduce the Banach space

$$\begin{aligned} \mathfrak{Y}^{l,q}_{\gamma}(\Omega) = \Big\{&(\sigma, \mathbf{v}) \in \mathfrak{V}^{(l,l+1),q}_{\gamma,2}(\Omega) \times \mathfrak{V}^{(l+1,l+2),q}_{\gamma,1}(\Omega)\colon \\ &\operatorname{div}\big((\rho_0+\sigma)\mathbf{v}\big) \in V^{l-1,q}_{\gamma}(\Omega),\ \mathbf{v}|_{\partial\Omega} = 0,\ (\operatorname{div}\mathbf{v})\big|_{\partial\Omega} = 0\Big\}, \end{aligned} \tag{6.5}$$

and let $B_{\varepsilon_{**}}$ be a ball in $\mathfrak{Y}^{l,q}_{\gamma}(\Omega)$:

$$B_{\varepsilon_{**}} = \Big\{(\sigma, \mathbf{v}) \in \mathfrak{Y}^{l,q}_{\gamma}(\Omega)\colon \big\|\sigma; \mathfrak{V}^{(l,l+1),q}_{\gamma,2}(\Omega)\big\| + \big\|\mathbf{v}; \mathfrak{V}^{(l+1,l+2),q}_{\gamma,1}(\Omega)\big\| \le \varepsilon_{**}\Big\}. \tag{6.6}$$

Assume that $(\tau, \mathbf{w}) \in B_{\varepsilon_{**}}$. Simple calculations, using Lemmata 2.1–2.5 and condition (6.3), furnish that

$$\mathbf{F}(\tau, \mathbf{w}) = (\rho_0+\tau)(\mathbf{w}\cdot\nabla)\mathbf{w} + (\rho_0+\tau)\mathbf{f} \in \mathfrak{V}^{(l-1,l),q}_{\gamma,3}(\Omega) \tag{6.7}$$

and

$$\begin{aligned} &\big\|\mathbf{F}(\tau, \mathbf{w}); \mathfrak{V}^{(l-1,l),q}_{\gamma,3}(\Omega)\big\| \\ &\quad\le c\Big(1 + \big\|\tau; \mathfrak{V}^{(l,l+1),q}_{\gamma,2}(\Omega)\big\|\Big)\Big(\big\|\mathbf{w}; \mathfrak{V}^{(l+1,l+2),q}_{\gamma,1}(\Omega)\big\|^2 + \big\|\mathbf{f}; \mathfrak{V}^{(l-1,l),q}_{\gamma,3}(\Omega)\big\|\Big) \\ &\quad\le c(\varepsilon_* + \varepsilon_{**}(1+\varepsilon_{**}))\Big(\big\|\mathbf{w}; \mathfrak{V}^{(l+1,l+2),q}_{\gamma,1}(\Omega)\big\| + \big\|\tau; \mathfrak{V}^{(l,l+1),q}_{\gamma,2}(\Omega)\big\|\Big) + c\varepsilon_*. \end{aligned}$$

Moreover, $\mathbf{F}(\tau, \mathbf{w})$ can be represented in the form

$$\mathbf{F}(\tau, \mathbf{w}) = r^{-3}\mathfrak{F}_1(\theta) + \widetilde{\mathbf{F}}(x),$$

where $\widetilde{\mathbf{F}} \in V^{l-1,q}_{\gamma}(\Omega)$ and

$$\frac{1}{r^3}\mathfrak{F}_1(\theta) = \operatorname{div}\big(\rho_* r^{-2}\mathfrak{W}(\theta)\otimes\mathfrak{W}(\theta)\big) + \rho_* r^{-3}\mathfrak{F}(\theta)$$

($\mathfrak{F}(\theta)$ and $\mathfrak{W}(\theta)$ are the spherical attributes of $\mathbf{f}$ and $\mathbf{w}$, respectively). Therefore, by Lemma 6.1, $\mathbf{F}(\tau, \mathbf{w})$ satisfies compatibility condition (3.8). If $\varepsilon_{**}$ is sufficiently small (see condition (5.21)), then by Theorem 5.1 linearized problem (1.9) with $\mathbf{F} = \mathbf{F}(\tau, \mathbf{w})$ has a unique solution $(\sigma, \mathbf{v}) \in \mathfrak{V}^{(l,l+1)}_{\gamma,2}(\Omega) \times \mathfrak{V}^{(l+1,l+2),q}_{\gamma,1}(\Omega)$. From $(1.9)_2$ we get

$$\begin{aligned} &\operatorname{div}(\rho_0\mathbf{v}) = -\operatorname{div}(\sigma\mathbf{w}) \in V^{l-1,q}_{\gamma}(\Omega), \\ &(\operatorname{div}\mathbf{v})|_{\partial\Omega} = -\rho_0^{-1}(\nabla\rho_0\cdot\mathbf{v} + \nabla\sigma\cdot\mathbf{w} + \sigma\operatorname{div}\mathbf{w})|_{\partial\Omega} = 0. \end{aligned}$$

Thus, the map $\mathcal{N}$ is well defined on the space $\mathfrak{Y}^{l,q}_{\gamma}(\Omega)$. Moreover, estimates (5.3), (6.7) yield

$$\begin{aligned}&\|\sigma;\mathfrak{V}^{(l,l+1),q}_{\gamma,2}(\Omega)\| + \|\mathbf{v};\mathfrak{V}^{(l+1,l+2),q}_{\gamma,1}(\Omega)\|\\ &\quad\le c_3\big(\varepsilon_* + (\varepsilon_{**}(1+\varepsilon_{**})\big)\Big(\|\mathbf{w};\mathfrak{V}^{(l+1,l+2),q}_{\gamma,1}(\Omega)\| + \|\tau;\mathfrak{V}^{(l,l+1),q}_{\gamma,2}(\Omega)\|\Big) + c_3\varepsilon_*.\end{aligned}$$

This gives the inclusion

$$\mathcal{N}B_{\varepsilon_{**}} \subset B_{\varepsilon_{**}} \text{ for } 2c_3\varepsilon_* < \varepsilon_{**} \quad \text{and} \quad 2c_3\big(\varepsilon_* + \varepsilon_{**}(1+\varepsilon_{**})\big) < 1. \tag{6.8}$$

It is easy to verify that $B_{\varepsilon_{**}}$ is a closed subset of the Banach space

$$\mathfrak{X}^{l,q}_{\gamma}(\Omega) = \big\{(\sigma,\mathbf{v}) \in \mathfrak{V}^{(l-1,l),q}_{\gamma-1,2}(\Omega) \times \mathfrak{V}^{(l,l+1),q}_{\gamma-1,1}(\Omega)\colon \mathbf{v}|_{\partial\Omega} = 0\big\} \supset \mathfrak{Y}^{l,q}_{\gamma}(\Omega).$$

We prove that $\mathcal{N}$ is a contraction in $B_{\varepsilon_{**}}$ with respect of the weaker topology of the space $\mathfrak{X}^{l,q}_{\gamma}(\Omega)$. Denote

$$\begin{aligned}&(\sigma,\mathbf{v}) = \mathcal{N}(\tau,\mathbf{w}), \quad (\sigma_1,\mathbf{v}_1) = \mathcal{N}(\tau_1,\mathbf{w}_1),\\ &\widehat{\sigma} = \sigma - \sigma_1, \quad \widehat{\mathbf{v}} = \mathbf{v} - \mathbf{v}_1, \quad \widehat{\tau} = \tau - \tau_1, \quad \widehat{\mathbf{w}} = \mathbf{w} - \mathbf{w}_1,\\ &\widehat{\mathbf{F}} = \mathbf{F}(\tau,\mathbf{w}) - \mathbf{F}(\tau_1,\mathbf{w}_1)\\ &\quad = -(\rho_0+\tau)(\mathbf{w}\cdot\nabla)\mathbf{w} + (\rho_0+\tau_1)(\mathbf{w_1}\cdot\nabla)\mathbf{w_1} + \widehat{\tau}\mathbf{f}.\end{aligned}$$

In virtue of (1.9) the couple $(\widehat{\sigma},\widehat{\mathbf{v}})$ satisfies the following equations

$$\begin{aligned}&-\mu_1\Delta\widehat{\mathbf{v}} - (\mu_1+\mu_2)\nabla\operatorname{div}\widehat{\mathbf{v}} + \rho_0\nabla(\widehat{\sigma}/\rho_0) = \widehat{\mathbf{F}}, \quad x\in\Omega,\\ &\operatorname{div}(\rho_0\widehat{\mathbf{v}}) + \operatorname{div}(\mathbf{w}\widehat{\sigma}) = -\operatorname{div}(\sigma\widehat{\mathbf{w}}) \equiv \widehat{g}, \quad x\in\Omega,\\ &\widehat{\mathbf{v}} = 0, \quad x\in\partial\Omega.\end{aligned} \tag{6.9}$$

For $(\sigma,\mathbf{v})$, $(\sigma_1,\mathbf{v_1}) \in B_{\varepsilon_{**}}$ and $\|\mathbf{f};\mathfrak{V}^{(l,l+1),q}_{\gamma,3}(\Omega)\| < \varepsilon_*$, we find the estimates

$$\begin{aligned}&\|\widehat{g};V^{l-1,q}_{\gamma-1,2}(\Omega)\| \le c\,\varepsilon_{**}\|\widehat{\mathbf{w}};\mathfrak{V}^{(l,l+1),q}_{\gamma-1,1}(\Omega)\|,\\ &\|\widehat{\mathbf{F}};\mathfrak{V}^{(l-2,l-1),q}_{\gamma-1,3}(\Omega)\| \le c\,(\varepsilon_*+\varepsilon_{**})\Big(\|\widehat{\tau};\mathfrak{V}^{(l-1,l),q}_{\gamma-1,2}(\Omega)\| + \|\widehat{\mathbf{w}};\mathfrak{V}^{(l,l+1),q}_{\gamma-1,1}(\Omega)\|\Big).\end{aligned} \tag{6.10}$$

Thus, Theorem 5.1 (with $l-1$ instead of $l$ and $\gamma-1$ instead of $\gamma$) applied to (6.10) yields

$$\begin{aligned}&\|\widehat{\sigma};\mathfrak{V}^{(l-1,l),q}_{\gamma-1,2}(\Omega)\| + \|\widehat{\mathbf{v}};\mathfrak{V}^{(l,l+1),q}_{\gamma-1,1}(\Omega)\|\\ &\quad\le c_4(\varepsilon_*+\varepsilon_{**})\Big(\|\widehat{\tau};\mathfrak{V}^{(l-1,l),q}_{\gamma-1,2}(\Omega)\| + \|\widehat{\mathbf{w}};\mathfrak{V}^{(l,l+1),q}_{\gamma-1,1}(\Omega)\|\Big).\end{aligned}$$

This furnishes the contraction for $\mathcal{N}$ in the Banach space $\mathfrak{X}^{l,q}_{\gamma}(\Omega)$ provided that

$$c_4(\varepsilon_*+\varepsilon_{**}) < 1. \tag{6.11}$$

Therefore, if $\varepsilon_*$, $\varepsilon_{**}$ satisfy conditions (5.21), (6.9), (6.11), there exists a fixed point $(\sigma,\mathbf{v}) \in \mathfrak{Y}^{l,q}_{\gamma}(\Omega)$ of $\mathcal{N}$ which solves the (1.7), (1.8). This completes the proof. □

Analogous results are valid in the scale of weighted Hölder spaces with detached asymptotics.

**Theorem 6.3.** *Let* $\Omega$, $\Phi$ *and* $\rho_0$ *be as in Theorem* 6.2. *Suppose that* $\mathbf{f} \in \mathfrak{C}^{(l-1,l),\delta}_{\gamma,3}(\Omega)$, $l \geq 2$, $\delta \in (0,1)$, $\gamma \in (l+2+\delta,\, l+3+\delta)$, $\gamma_0 > 1$. *Assume also that* $\mathbf{f}$ *satisfies compatibility condition* (3.8). *There exists a number* $\varepsilon_* > 0$ *such that if*

$$\left\|\mathbf{f}; \mathfrak{C}^{(l-1,l),\delta}_{\gamma,3}(\Omega)\right\| < \varepsilon_*,$$

*then problem* (1.7), (1.8) *has exactly one solution*

$$(\sigma, \mathbf{v}) \in \mathfrak{C}^{(l,l+1),\delta}_{\gamma,2}(\Omega) \times \mathfrak{C}^{(l+1,l+2),\delta}_{\gamma,1}(\Omega)$$

*satisfying the estimate*

$$\left\|\sigma; \mathfrak{C}^{(l,l+1),\delta}_{\gamma,2}(\Omega)\right\| + \left\|\mathbf{v}; \mathfrak{C}^{(l+1,l+2),\delta}_{\gamma,1}(\Omega)\right\| \leq c \left\|\mathbf{f}; \mathfrak{C}^{(l-1,l),\delta}_{\gamma,3}(\Omega)\right\|. \tag{6.12}$$

*Remark* 6.4. Note that constants in estimates (6.4), (6.12) depend on $\left\|\Phi; \Lambda^{l+1,\delta}_{l+1+\delta+\gamma_0}(\Omega)\right\|$.

*Remark* 6.5. From the obtained results it follows, in particular, that in the case where $\mathbf{f}$ has a compact support the solution $(\sigma, \mathbf{v})$ admits the asymptotic representation

$$\sigma(x) = \frac{1}{r^2}\Sigma(\theta) + \widetilde{\sigma}(x), \quad \mathbf{v}(x) = \frac{1}{r}\mathbf{V}(\theta) + \widetilde{\mathbf{v}}(x),$$

with

$$\left|D^\alpha \widetilde{\mathbf{v}}(x)\right| = O(r^{-1-|\alpha|-\widehat{\gamma}}), \quad |\alpha| = 0, 1, \dots, l+1,$$
$$\left|D^\alpha \widetilde{\mathbf{p}}(x)\right| = O(r^{-2-|\alpha|-\widehat{\gamma}}), \quad |\alpha| = 0, 1, \dots, l, \quad \forall \widehat{\gamma} < 1.$$

## References

[1] W. Borchers and K. Pileckas. Existence, uniqueness and asymptotics of steady jets. *Arch. Rat. Mech. Analysis* **120** (1983), 1–49.

[2] A. Matsumura and T. Nishida. Initial boundary value problems for equations of motion of compressible viscous and heat-conductive fluids. *Comm. Math. Phys.* **89** (1983), 445–464.

[3] S.A. Nazarov. On the two-dimensional aperture problem for Navier–Stokes equations. *C.R. Acad. Sci. Paris, Ser. 1* **323** (1996), 699–703.

[4] S.A. Nazarov. The Navier–Stokes problem in a two-dimensional domains with angular outlets to infinity. *Zapiski Nauchn. Seminarov POMI* **257** (1999), 207–227 (in Russian).

[5] S.A. Nazarov. Weighted spaces with detached asymptotics in application to the Navier–Stokes equations. *Advances in Math. Fluid Mechanics, Lecture Notes of the Sixth International School "Mathematical Theory in Fluid Mechanics"*, J. Malek, J. Nečas, M. Rokyta (Eds.), Springer, 2000, 159–191.

[6] S.A. Nazarov and K. Pileckas. On steady Stokes and Navier–Stokes problems with zero velocity at infinity in a three-dimensional exterior domain. *J. Math. Kyoto Univ.* **40**(3) (2000), 475–492.

[7] S.A. Nazarov and K. Pileckas. Asymptotics of solutions to the Navier–Stokes equations in the exterior of a bounded body. *Doklady RAN* **367**(4) (1999), 461–463. English transl.: *Doklady Math.* **60**(1) (1999), 133–135.

[8] S.A. Nazarov and K. Pileckas. Asymptotic conditions at infinity for the Stokes and Navier–Stokes problems in domains with cylindrical outlets to infinity. *Advances in Fluid Dynamics, Quaderni di Matematica,* P. Maremonti (Ed.) **4** (1999), 141–243.

[9] S.A. Nazarov and B.A. Plamenevskii. *Elliptic boundary value problems in domains with piecewise smooth boundaries.* Walter de Gruyter and Co, Berlin, 1994.

[10] S.A. Nazarov, A. Sequeira and J.H. Videman. Asymptotic behaviour at infinity of three-dimensional steady viscoelastic flows. *Pacific J. Math.* **203** (2002), 461–488.

[11] S.A. Nazarov, M. Specovius-Nengebaner and G. Thäter. Quiet flows for Stokes and Navier–Stokes problems in domains with cylindrical outlets to infinity. *Kyushu J. Math.* **53** (1999), 369–394.

[12] A. Novotny. On steady transport equation. *Advanced Topics in Theoretical Fluid Mechanics, Pitman Research Notes in Mathematics,* J. Malek, J. Nečas, M. Rokyta (Eds.), **392** (1998), 118–146.

[13] A. Novotny. About steady transport equation II, Shauder estimates in domains with smooth boundaries. *Portugaliae Mathematica,* **54**(3) (1997), 317–333.

[14] A. Novotny and M. Padula. Physically reasonable solutions to steady compressible Navier–Stokes equations in 3D-exterior domains I ($v_\infty = 0$). *J. Math. Kyoto Univ.* **36**(2) (1996), 389–423.

[15] A. Novotny and K. Pileckas. Steady compressible Navier–Stokes equations with large potential forces via a method of decomposition. *Math. Meth. in Appl. Sci.* **21** (1998), 665–684.

[16] V.A. Solonnikov. On the solvability of boundary and initial-boundary value problems for the Navier–Stokes system in domains with noncompact boundaries. *Pacific J. Math.* **93**(2) (1981), 443–458.

T. Leonavičienė and K. Pileckas
Institute of Mathematics and Informatics
Akademijos 4
2600 Vilnius
Lithuania
e-mail: `Terese.Brazauskaite@vpu.lt`
`pileckas@julius.ktl.mii.lt`

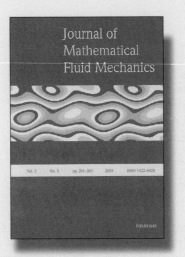